KW-171-470

RESPONSE TO MARINE OIL SPILLS

THE INTERNATIONAL TANKER OWNERS POLLUTION FEDERATION LTD

First published in 1986 as five booklets by
The International Tanker Owners Pollution Federation Ltd., London

This edition published in 1987 by
Witherby & Co. Ltd., London
Reprinted 1990

© *The International Tanker Owners Pollution Federation Ltd., London*

ISBN 0 948691 51 4

This book is intended to reflect the best available techniques and
practices at the time of publication. However, the authors and
publishers cannot be held responsible should any of the techniques and
practices prove unsuccessful in any particular instance. Each spill
situation will be different and the techniques to be employed will vary
according to the circumstances.

Printed in England by
Witherby & Co. Ltd., London, EC1

PREFACE

This book provides a comprehensive review of the problems posed by marine oil spills and the response measures that can be implemented. Emphasis throughout the book is on practical guidance, based on the experience gained by the authors through their involvement with marine oil spills around the world.

The individual sections of this book supplement information contained in a series of five oil spill training videos entitled "Response to Marine Oil Spills". The production of these videos was sponsored by the International Maritime Organization, The International Tanker Owners Pollution Federation, the Commission of the European Communities and Videotel Marine International. The five videos, which run for a total of about 100 minutes, plus the supporting books, are marketed by Videotel Marine International as a training package. English, French and Spanish language versions are currently available in various video formats and for various colour standards. Further information on the training package can be obtained from:

Videotel Marine International,
Ramillies House,
1/2 Ramillies Street,
London W1V 1DF,
UK.

Telephone: (01) 439 6301/5
Telex:　　298596 VIDTEL G

For further information on responding to marine oil spills contact:

The International Tanker Owners Pollution Federation Ltd.,
Staple Hall,
Stonehouse Court,
87-90 Houndsditch,
London EC3A 7AX,
UK.

Telephone: (01) 621 1255 (5 lines)
Telex:　　887514 TOVLOP G
Fax:　　　(01) 626 5913

(Note: With effect from 6th May, 1990 the 01 prefix in the above numbers will
　　　be replaced by 071)

CONTENTS

V. PLANNING AND OPERATIONS

ACKNOWLEDGEMENTS

I THE OIL SPILL

Oil enters the marine environment by a number of different routes as a result of both human activities and natural processes. Tanker accidents and offshore blow-outs account for about 15% of the total amount of oil entering the oceans.

This section examines the various sources of hydrocarbons in the marine environment in order to place accidental oil spills in perspective. The fate and effects of oil spills are also described and general advice is given on the monitoring of oil slicks and the quantification of pollution at sea and on shore.

CONTENTS

SOURCES OF PETROLEUM HYDROCARBONS

Major inputs

World oil demand

The best estimate for the total input of petroleum to the marine environment from all sources is some 3.2 million metric tonnes per year. By far the biggest contribution comes from terrestrial sources, mainly in the form of municipal and industrial wastes. Accidental spills from ships, together with offshore exploration and production activities, account for about 0.47 million metric tonnes which is a relatively small amount considering the world's current production of three billion metric tonnes, half of which is transported by sea (Figure 1).

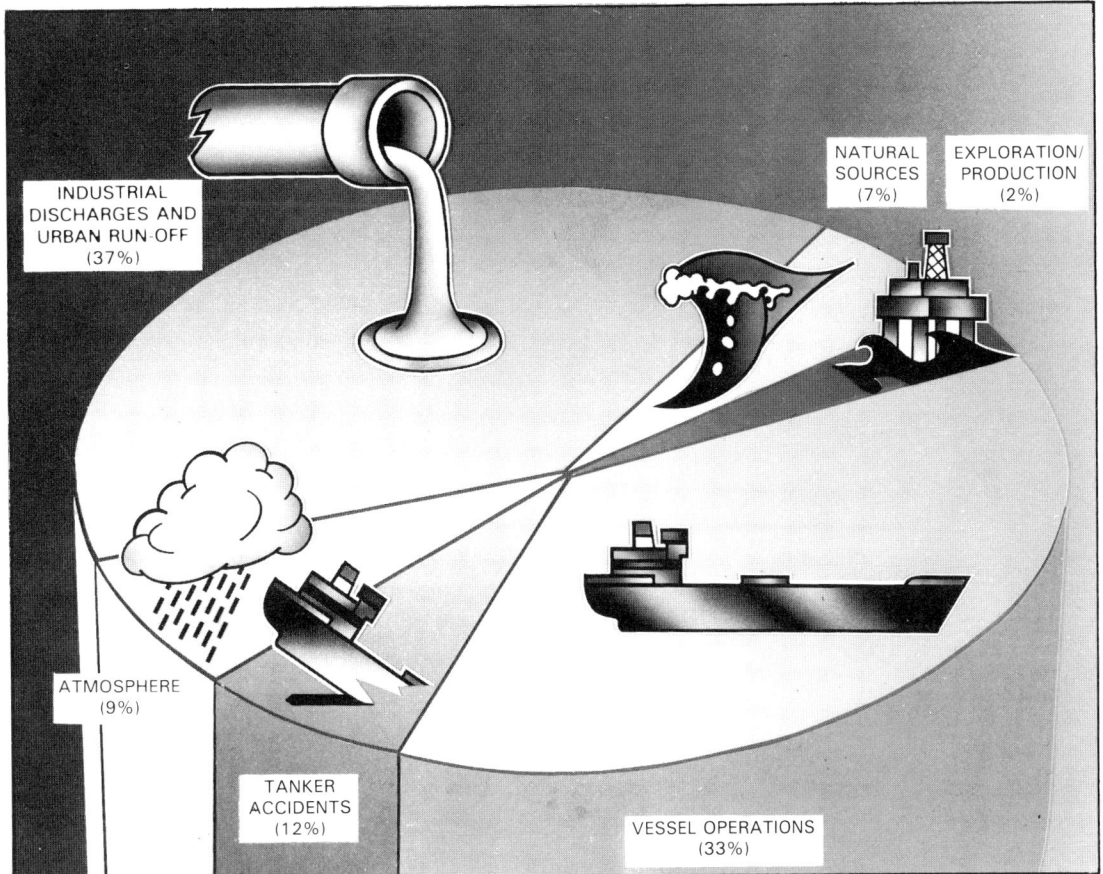

Figure 1:
Major inputs of petroleum to the marine environment.

Transportation Losses

Operational discharges

Limiting discharges

Of the total 1.47 million metric tonnes that enter the sea each year as a result of transportation losses, 0.7 million tonnes can be traced back to cargo residues remaining on board after discharge and referred to as clingage. The amount of clingage depends mainly on the wax content and viscosity of the previous cargo, but typically amounts to about 0.4% of the cargo carrying capacity, i.e. 800 tonnes on a 200,000 DWT crude carrier. During tank cleaning and de-ballasting operations, much of this can be lost overboard unless precautions are taken to retain oily slops on board. Recent developments such as Segregated Ballast Tank (SBT) arrangements and Crude Oil Washing (COW) systems, together with established 'Load On Top' (LOT) procedures, have reduced operational pollution from tankers. Included in transportation losses are discharges of oily bilge water and fuel oil sludges, amounting to 300,000 tonnes annually, which are generated by all types of ships. Although the quantities of waste oil discharged to the sea from ships can be controlled through strict management, great importance is also attached to the provision of adequate shore reception facilities for tanker slops, dirty ballast water and oily residues from machinery spaces.

Accidental Spills from Tankers

Routine operations

Major accidents

Accidental spills from tankers contribute an estimated 400,000 tonnes annually. Analysis of tanker spills occurring throughout the world shows that the majority (some 75%) occur in port during routine ship operations such as loading, discharging and bunkering. Most of these spills are, however, relatively small: over 92% are less than 7 tonnes (Table 1) and probably, in total, contribute less than 20,000 tonnes annually. In comparison, accidents such as collisions and groundings give rise to less than 10% of all spills from tankers, but a quarter of these are larger than 700 tonnes (Table 1). In fact, a few large accidents give rise to the majority of the oil spilt and hence there is considerable annual variation in this figure.

Table 1. Comparison of Incidence of World Oil Spills from Tankers, 1974-1985, resulting from Routine Operations and Major Accidents

	<7 (tonnes)	7-700 (tonnes)	>700 (tonnes)	Total
Loading/ discharging	2236 (90%)	227 (9%)	11 (1%)	2474 (100%)
Bunkering	442 (95%)	22 (5%)	—	464 (100%)
Collision	39 (17%)	134 (59%)	54 (24%)	227 (100%)
Grounding	69 (25%)	134 (49%)	70 (26%)	273 (100%)
Total	2786 (81%)	517 (15%)	135 (4%)	3438 (100%)

Offshore Oil Exploration and Production Activities

Blow-outs

Operational losses

Major pollution incidents, such as blow-outs, are rare but contribute roughly three quarters of the 50,000 tonnes lost annually from offshore platforms. The risk is less during production than in the exploration phase, but a blow-out can result in large volumes of oil being lost if the well is not brought under control quickly. A much larger number of small releases occur as a result of routine operations such as the discharge of formation water and disposal of oil-based drilling muds.

Terrestrial and Atmospheric Inputs

Industry and urban run-off

Vehicle exhausts

Terrestrial oil inputs are principally from discharges of process water from coastal refineries and other industries; waste oils carried to sea in sewage discharges and rivers; and urban run-off from road networks. By comparison, the atmospheric fallout of petroleum hydrocarbons is probably less significant, but the extent is very difficult to estimate accurately on a global scale. The main portion of marine pollution from the atmosphere can be linked to exhaust fumes from road vehicles.

Natural Seeps and Erosion

Sediment erosion

Natural inputs are also difficult to quantify and show a very patchy distribution. Seeps tend to be associated with regions of tectonic activity in oceanic margins whereas erosion of exposed oil-rich sediments will take place at terrestrial locations and usually form part of the river run-off.

FATE OF MARINE OIL SPILLS

Natural assimilation

Despite the introduction of many millions of tonnes of oil into the world's oceans, there is little evidence of a build-up of oil residues in the sea. This is a good indication that the marine environment is able to assimilate oil. Accidental oil spills are usually of greatest concern since these often give rise to conspicuous acute pollution.

Persistence of oil at sea

Oil spilled into the sea undergoes a number of physical and chemical changes, some of which lead to its disappearance from the sea surface whilst others cause it to persist. The time taken depends primarily upon the physical and chemical characteristics of the oil, as well as the quantity involved, the prevailing climatic and sea conditions and whether the oil remains at sea or is washed ashore.

Properties of Oil

Oil types

In considering the fate of spilled oil at sea, a distinction is frequently made between non-persistent oils, which tend to disappear rapidly from the sea surface, and persistent oils, which in contrast, dissipate more slowly and usually require a clean-up response. Non-persistent oils include gasolene, naphtha, kerosene and diesel whereas most crude oils and refined residual oils have varying degrees of persistence depending on their physical properties and the size of the spill.

The main physical properties which affect the behaviour of an oil spilled at sea are specific gravity, distillation characteristics, viscosity and pour point.

Specific gravity

The specific gravity of an oil is its density in relation to pure water. Most oils are lighter than water and have a specific gravity below 1. The density of crude oils and petroleum products is usually expressed in terms of API gravity in accordance with the following formula:

°API

$$°API = \frac{141.5}{\text{Specific Gravity}} - 131.5$$

In addition to determining whether or not the oil will float, its density can also give a general indication of other properties of the oil. For example, oils with a low specific gravity (high °API) tend to be rich in volatile components and highly fluid.

Distillation characteristics

The distillation characteristics of an oil describe its volatility. As the temperature of an oil is raised, different components reach their boiling point in turn and are distilled. The distillation characteristics are expressed as the proportions of the parent oil which distill within given temperature ranges.

Viscosity

The viscosity of an oil is its resistance to flow. High viscosity oils flow with difficulty whilst those with low viscosities are highly fluid. Viscosities decrease at higher temperatures and so sea water temperature and the extent to which the oil can absorb heat from the sun are important considerations.

Pour point

The pour point is the temperature below which an oil will not flow. If the ambient temperature is below the pour point, the oil will essentially behave as a solid.

Weathering Processes

Need for spill response

The physical and chemical changes which spilled oil undergo are sometimes collectively known as weathering. The various processes are shown schematically in Figure 2. A knowledge of these processes and how they interact to alter the nature and composition of the oil with time is valuable in preparing and implementing contingency plans for oil spill response. On occasions it may prove unnecessary to mount a clean-up response if it can be confidently predicted that the oil will drift away from vulnerable resources or dissipate naturally before reaching them. Often, however, an active response will be necessary, aimed either at accelerating the natural processes through the use of dispersants or limiting spreading by containment.

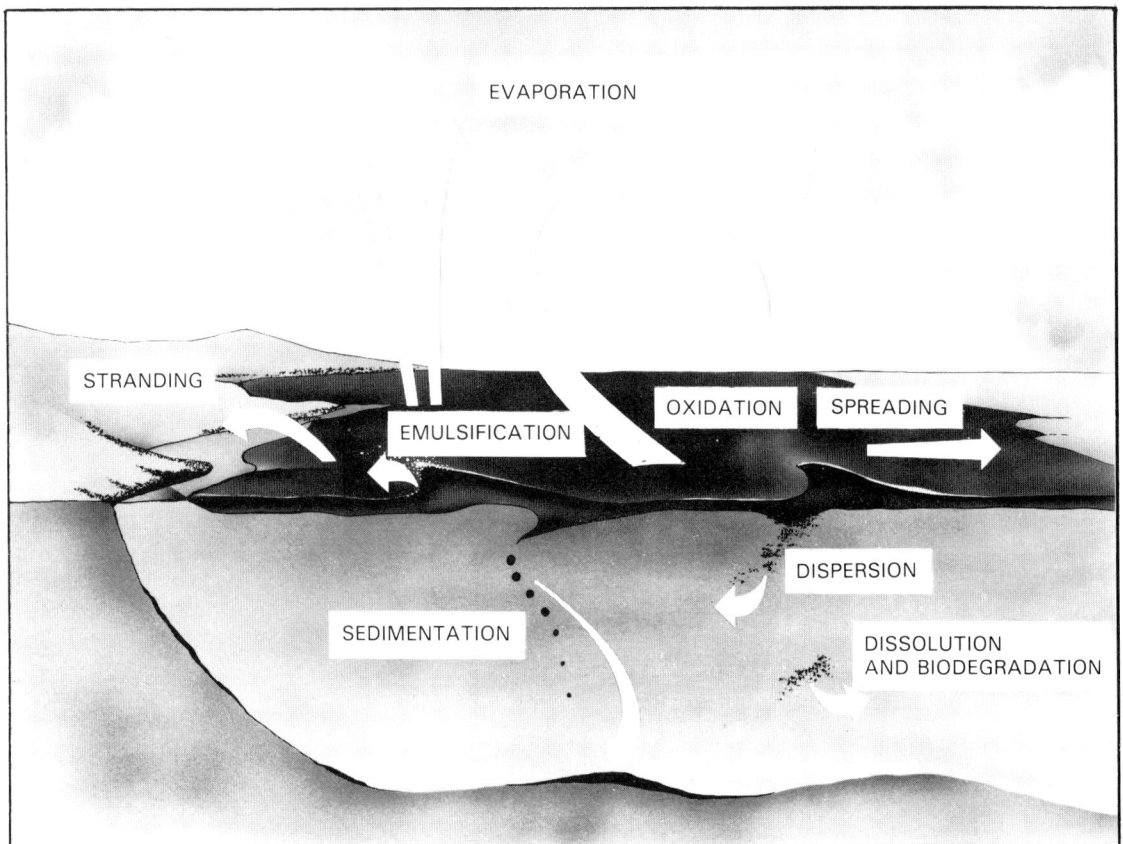

Figure 2:
Fate of spilt oil including the main weathering processes.

Spreading

Initial spreading

Spreading is one of the most significant processes during the early stages of a spill. The main driving force behind the initial spreading of the oil is its weight. A large instantaneous spill will therefore spread more rapidly than a slow discharge. This gravity-assisted spreading is quickly replaced by surface tension effects. During these early stages, the oil spreads as a coherent slick and the rate is also influenced by the viscosity of the oil. High viscosity oils spread only slowly and those spilled at temperatures below their pour point hardly spread at all. After a few hours the slick begins to break up and form narrow

Windrows

bands or 'windrows' parallel to the wind direction. At this stage the fluidity of the oil becomes less important since further spreading is primarily due to turbulence at the sea surface. Variations in spreading rate are due to differences in the prevailing hydrographical

Scattering

conditions such as currents, tidal streams and wind speeds. Some 12 hours after a spill, the oil can be scattered within an area of up to 5 square kilometres thus limiting the possibility of effective clean-up of oil at sea. It should be appreciated that, except in the case of small spills of low viscosity oils, spreading is not uniform and large variations of oil thickness occur within the slick.

Evaporation

Influencing factors

The rate and extent of evaporation is determined primarily by the volatility of the oil. The greater the proportion of components with low boiling points, the greater the evaporation. The initial spreading rate of the oil also affects evaporation since the larger the surface area, the faster the light components will evaporate. Rough seas, high wind speeds and warm temperatures will further increase the rate of evaporation. In broad terms, those oil components with a boiling point below 200°C will evaporate within a period of 24 hours in temperate conditions.

Evaporation rates

Spills of refined products, such as kerosene and gasolene, may evaporate completely within a few hours and light crudes can lose up to 40% during the first day. In contrast, heavy crudes and fuel oils undergo little, if any, evaporation. Any residue of an oil remaining after evaporation will have an increased density and viscosity which affects further weathering processes as well as the choice of clean-up techniques.

Fire and explosion hazard

When extremely volatile oils are spilled in confined areas, there may be a risk of fire and explosion. The flammability of oil has often led to the idea of burning slicks on the sea surface. Although it is often possible to ignite slicks, particularly of fresh oil, it is difficult to maintain combustion even when wicking agents are employed due to the thinness of the oil layer and the cooling effect of the water underneath. The residues remaining after partial combustion are usually more troublesome and difficult to deal with than naturally weathered oil.

Dispersion

Oil droplet behaviour

Waves and turbulence at the sea surface act on the slick to produce oil droplets with a range of sizes. Small droplets remain in suspension while the larger ones rise back to the surface, behind the advancing slick, where they may either coalesce with other droplets to reform a slick, or spread out in a very thin film. Droplets small enough to remain in suspension become mixed into the water column and the increased surface area presented by this dispersed oil can enhance other processes such as biodegradation and sedimentation.

Influence of sea state

The rate of natural dispersion is largely dependent upon the nature of the oil and the sea state, proceeding most quickly in the presence of breaking waves. Slick thickness, which is related to the amount spilled and the degree of spreading, is an important factor in the rate of dispersion since smaller droplets are produced from thin films.

Influence of viscosity

Oils which remain fluid and can spread unhindered by other weathering processes may disperse completely in moderate sea conditions within a few days. Conversely, viscous oils or those which form stable water-in-oil emulsions tend to form thick lenses on the water surface, and will show little tendency to disperse. Such oils can persist for several weeks.

Emulsification

Persistence

'Mousse'

Many oils tend to absorb water to form water-in-oil emulsions which can increase the volume of pollutant by a factor of between three and four. Such emulsions are often extremely viscous and so the other processes which would dissipate the oil are retarded. This is the main reason for the persistence of light and medium crude oils on the sea surface. In moderate to rough sea conditions most oils rapidly form emulsions, the stability of which are dependent on the concentration of asphaltenes. Oils with asphaltene contents greater than 0.5% tend to form stable emulsions, often referred to as "chocolate mousse", whilst those containing less are likely to disperse. Emulsions may separate out into oil and water again if heated by sunlight under calm conditions or when stranded on shorelines.

Water content

The rate at which emulsification takes place is primarily a function of sea state although viscous oils tend to absorb water more slowly. In wind strengths greater than about Beaufort Force 3, some low viscosity oils can incorporate between 60% and 80% water by volume within about 2-3 hours. In contrast, very viscous oils may take 10 hours or more to absorb 10% water under the same conditions and even after several days the water content seldom exceeds 40%.

Change in properties

Absorption of water usually results in black oil changing colour to brown, orange or yellow. As the emulsion develops, the movement of the oil in the waves causes droplets of water taken up in the oil to become smaller and smaller making the emulsion progressively more viscous. As the amount of water absorbed increases, the density of the emulsion approaches that of sea water.

Dissolution

Soluble components

The rate and extent to which an oil dissolves depends upon its composition, extent of spreading, water temperature, turbulence and degree of dispersion. The heavy components of crude oil are virtually insoluble in sea water whereas lighter compounds, particularly aromatic hydrocarbons such as benzene and toluene, are slightly soluble. However, these components are also the most volatile and so are lost very rapidly by evaporation, typically 10-1000 times faster than by dissolution. Concentrations of dissolved hydrocarbons thus rarely exceed one part per million and dissolution does not make a significant contribution to the removal of oil from the sea surface.

Oxidation

Reaction with oxygen

Sunlight

Tar balls

Hydrocarbon molecules react with oxygen and either break down into soluble products or combine to form persistent tars. Many of these oxidation reactions are promoted by sunlight and although they occur throughout the lifetime of a slick, the effect on the overall dissipation is minor in relation to other weathering processes. Under strong sunlight, thin films break down at rates of no more than 0.1% per day. Oxidation of thick layers of high viscosity oils or water-in-oil emulsions is more likely to lead to their persistence than to their degradation. This is due to the formation of higher molecular weight compounds which form an outer protective skin. For example, the tarry deposits which sometimes strand on shorelines as tar balls usually consist of a solid outer crust combined with sediment particles surrounding a softer, less weathered interior.

Sedimentation

Heavy crudes

Temperature effect

Some heavy residual oils have specific gravities greater than 1, and so will sink in fresh or brackish water. However, very few crude oils are sufficiently dense, or weather to such an extent that the residues alone will sink in sea water. Sinking is usually brought about by adhesion of particles of sediment or organic matter to the oil. Some heavy crudes, such as those produced in Venezuela, as well as most heavy fuel oils and water-in-oil emulsions have specific gravities near 1, and therefore require very little particulate matter to exceed the specific gravity of sea water (about 1.025). Temperature can also be expected to affect the behaviour of neutrally buoyant oil. Over a 10°C temperature range the density of sea water will only change by 0.25% whereas oil density changes by 0.5%. Oil which barely floats during the day may therefore submerge as the temperature falls at night due to its greater relative increase in density but may resurface later in warmer water.

Mechanisms of sediment-ation

Shallow waters are often laden with suspended solids providing favourable conditions for sedimentation. It is less likely in the open sea, but zooplankton may inadvertently take in particles of oil during feeding which become incorporated into faecal pellets which fall to the seabed.

Tar mats

Oil stranded on sandy shorelines often becomes mixed with sediments and if this mixture is subsequently washed off the beach it may sink. On exposed sand beaches, heavy contamination may lead to accumulation of large amounts of sediment in the oil, forming dense tar mats. Seasonal cycles of sediment build-up and erosion may cause oil layers to be successively buried and uncovered. Sheltered shorelines tend to be made up of fine grained sediments and if oil becomes incorporated in these, it is likely to remain there for a considerable time.

Biodegradation

Micro-organisms

Oxygen and nutrients

Sea water contains a range of marine bacteria, moulds and yeasts which can utilise oil as a source of carbon and energy. Such micro-organisms are widely distributed in the sea although they tend to be more abundant in chronically polluted waters, such as those which receive industrial discharges and untreated sewage. The main factors affecting the rate of biodegradation are temperature and the availability of oxygen and nutrients, principally compounds of nitrogen and phosphorous. Each type of micro-organism tends to degrade a specific group of hydrocarbons and whilst a range of bacteria exists which between them are capable of degrading most of the wide variety of compounds in crude oil, some components are resistant to attack. Although the micro-organisms are not always present in sufficient numbers in the open sea, with the right conditions they multiply rapidly until the process becomes limited by nutrient or oxygen deficiency.

Oil/water interface

Rates of biodegrad-ation

Because the micro-organisms live in seawater, biodegradation can only take place at an oil/water interface. Oil stranded on shorelines above high water mark will therefore break down extremely slowly and may persist for many years. At sea, the creation of oil droplets, either through natural or chemical dispersion, increases the interfacial area available for biological activity and so enhances degradation. The variety of factors influencing biodegradation makes it difficult to predict the rate of oil removal. In temperate waters daily rates of between 0.001 and 0.03 grams per tonne of sea water have been reported but may reach 0.5-60 grams per tonne of sea water in areas chronically polluted by oil. Once oils become incorporated into sediments, however, degradation rates are very much reduced due to a lack of oxygen and nutrients.

Combined Processes

The processes of spreading, evaporation, dispersion, emulsification and dissolution are most important during the early stages of a spill, whilst oxidation, sedimentation and biodegradation are long-term processes which determine the ultimate fate of oil.

Classifi-cation of oils

Wax content

Since the mechanisms of interaction between the various weathering processes are not well understood, reliance is often placed on empirical models based upon oil type. For this purpose it is convenient to classify the most commonly transported oils into four main groups roughly according to their specific gravity (Table 2, overleaf). As a general rule, the lower the specific gravity of the oil, the less persistent it will be. However, it is important to appreciate that some apparently light oils behave more like heavy ones due to the presence of waxes. Oils with wax contents greater than about 10% tend to have high pour points and if the ambient temperature is below this, the oil will behave either as a solid or as a highly viscous liquid.

Half-life concept

Having classified the oils, one way in which their persistence can be described is in terms of a half-life for each group. This is the time taken for the removal of 50% of the oil from the sea surface. After six half-lives, little more than 1% of the oil will remain. This model is shown in Figure 3, overleaf, which also takes into account the effect of emulsification on the volume of oil over time. The half-lives have been selected on the basis of observations made in the field and are intended to give an impression of how persistence varies according to the physical properties of the oil. In the same way that

Table 2. CLASSIFICATION OF COMMON CRUDE AND FUEL OILS ACCORDING TO THEIR SPECIFIC GRAVITY

Group I

Specific Gravity **<0.8** (°API **> 45**)

B Viscosity cSt @ 15°C: **0.5 — 2.0**
C % boiling below 200°C: **50 — 100%**
D % boiling above 370°C: **0%**

	B	C	D
Gasolene	0.5	100	0
Naptha	0.5	100	0
Kerosene	2.0	50	0

Group II

Specific Gravity **0.8 — 0.85** (°API **35 — 45**)

A Pour point °C
B Viscosity cSt @ 15°C: **4 — solid Average 8 cSt**
C % boiling below 200°C: **10 — 48% Average 35%**
D % boiling above 370°C: **0 — 40% Average 30%**

High pour point >5°C*	A	B	C	D	Low pour point	B	C	D
Argyll	9	11	29	39	Abu Dhabi	7	36	31
Amna	18	s	25	30	Berri	9	36	35
Arjuna	27	s	37	15	Beryl	9	35	34
Auk	9	9	33	35	Brass River	4	45	17
Bass Straight	15	s	40	20	Brega	9	38	32
Beatrice	12	32	25	35	Brent Spar	9	37	32
Bunyu	18	s	29	12	Ekofisk	4	46	25
Cormorant	12	13	32	38	Kirkuk	11	35	36
Dunlin	6	11	29	36	Kole Marine	11	34	35
Escravos	10	9	35	15	Montrose	7	36	31
Es Sider	9	11	31	37	Murban	9	36	30
Gippsland Mix	15	s	40	20	Murchison	7	36	20
Lucina	15	s	30	34	Qatar Marine	9	36	33
Nigerian Light	9	s	35	27	Saharan Blend	4	48	23
Ninian	6	13	29	40	Sirtica	7	44	27
Qua Iboe	10	7	37	8	Statfjord	9	35	32
Rio Zulia	27	s	34	30	Zakum	7	34	31
San Joachim	24	s	43	20				
Santa Rosa	10	4	34	27	Gas Oil	5	—	—
Sarir	24	s	24	39				
Seria	18	s	37	15				
Thistle	9	9	35	38				
Zuetina	9	9	35	30				

*These oils would only behave as group 2 at ambient temperatures above their pour points. At lower temperatures treat as group 4 oils.

Group III

Specific Gravity **0.85 — 0.95** (°API **17.5 — 35**)

A Pour point °C
B Viscosity cSt @ 15°C: **8 — solid Average 275 cSt**
C % boiling below 200°C: **14 — 34% Average 25%**
D % boiling above 370°C: **28 — 60% Average 45%**

High pour point >5°C*	A	B	C	D	Low pour point	B	C	D
Bakr	7	1500	14	60	Arabian Light	14	30	40
Belayim (marine)	15	s	22	55	Arabian Medium	25	29	45
Cabinda	21	s	21	52	Arabian Heavy	55	25	49
El Morgan	7	30	25	47	Buchan	14	31	39
Mandji	9	70	21	53	Champion Export	18	15	26
Soyo	15	s	21	48	Flotta	11	34	26
Suez Mix	10	30	24	49	Forcados	12	18	34
Trinidad	14	s	23	28	Forties	8	32	36
Zaire	15	s	18	55	Iranian Heavy	25	29	44
					Khafji	80	25	49
					Kuwait	30	29	46
					Maya	500	25	49
					Nigerian Medium	40	14	40
					Santa Maria	250	22	54
					Tia Juana Light	2500	24	45
					Medium Fuel Oil	1500-3000	—	—

*These oils would only behave as group 3 at ambient temperatures above their pour points. At lower temperatures treat as group 4 oils.

Group IV

Specific Gravity **> 0.95** (°API **< 17.5**)
or Pour Point **> 30°C**

A Pour point °C
B Viscosity cSt @ 15°C: **1500 — solid**
C % boiling below 200°C: **3 — 24% Average 10%**
D % boiling above 370°C: **33 — 92% Average 65%**

	A	B	C	D
Bachequero Heavy	− 20	5,000	10	60
Bahia	38	s	24	45
Boscan	15	s	4	80
Bu Attifil	39	s	19	47
Cinta	43	s	10	54
Cyrus	− 12	10,000	12	66
Duri	14	s	5	74
Gamba	23	s	11	54
Handil	35	s	23	33
Heavy Lake Mix	− 12	10,000	12	64
Jatibarang	43	s	14	65
Jobo/Morichal	− 1	23,000	3	76
Lagunillas	− 20	7,000	9	73
Merey	− 23	7,000	10	66
Minas	36	s	17	53
Panuco	2	s	3	76
Pilon	− 4	s	2	92
Quiriquire	− 29	1,500	3	88
Shengli	21	s	9	70
Taching	35	s	12	49
Tia Juana Pesado	− 1	s	3	78
Wafra Eocene	− 29	3,000	11	63
Heavy Fuel Oil (Bunker C)		5,000-30,000	—	—

Climate and weather

no single oil has the exact properties of those indicated by the groups given in Table 2, weather and climatic conditions will also influence the half life of a slick. For example, in very rough weather, an oil in Group 3 may dissipate within a timescale more typical of a Group 2 oil. Conversely, in cold calm conditions it may approach the persistence of Group 4 oils.

Response decisions

It is important to appreciate the assumptions made with models such as the one described and not to place too much reliance on the results. However, they can provide a guide to which clean-up techniques are likely to be effective, whether a response can be initiated quickly enough, and what sort of problem a clean-up organisation will have to face.

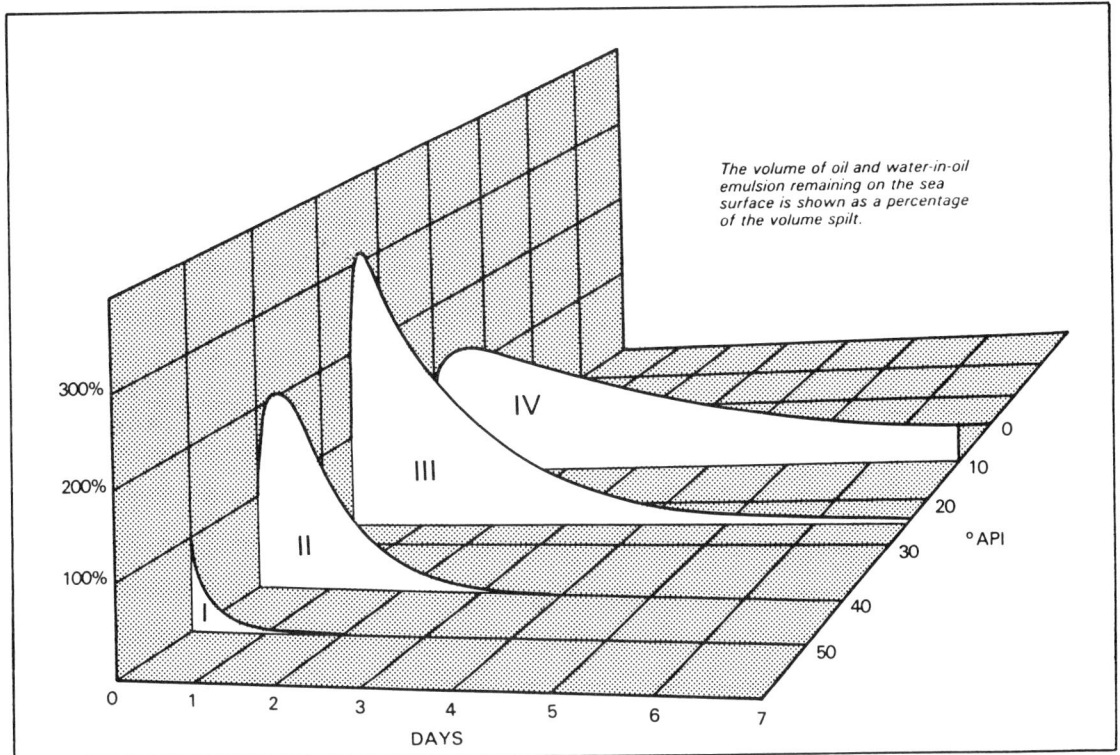

The volume of oil and water-in-oil emulsion remaining on the sea surface is shown as a percentage of the volume spilt.

Figure 3:
The rate of removal of oil groups I-IV from the sea surface according to their physical properties.

Forecasting Slick Movement

It is equally important to be able to forecast the probable movement of a slick as well as the likely changes in the properties of an oil after it has been spilled. This allows sensitive resources in the path of the slick to be identified and, if appropriate, response measures to be put into effect. The task of forecasting the position of the oil can only be accomplished if data on winds and currents are available since both contribute to the movement of floating oil.

Effect of wind

Tidal currents

It has been found empirically that floating oil will move downwind at about 3% of the wind speed. In the presence of surface water currents, an additional movement of the oil equivalent to the current strength will be superimposed on any wind-driven motion. Close to land, the strength and direction of any tidal currents must be taken into account but further out to sea their contribution is usually less significant because they are cyclic and so tend to cancel out over time. Thus, with a knowledge of the prevailing winds and currents, it is possible to predict the rate and direction of movement of floating oil from a known position, as shown in Figure 4, overleaf.

This simple calculation can be easily done by hand but becomes very time-consuming if tidal currents have to be taken into account since it must be recalculated at regular intervals as currents change. Computers can be used to speed up such calculations by storing information on water movement and coastal outline for a specific geographic area. Wind data and the spill location are then the only additional information required at the time of a spill. The reliability of such models depends upon the accuracy of water movement and wind data. Often they are combined with mathematical models simulating weathering processes to provide a forecast of the overall fate of a spill.

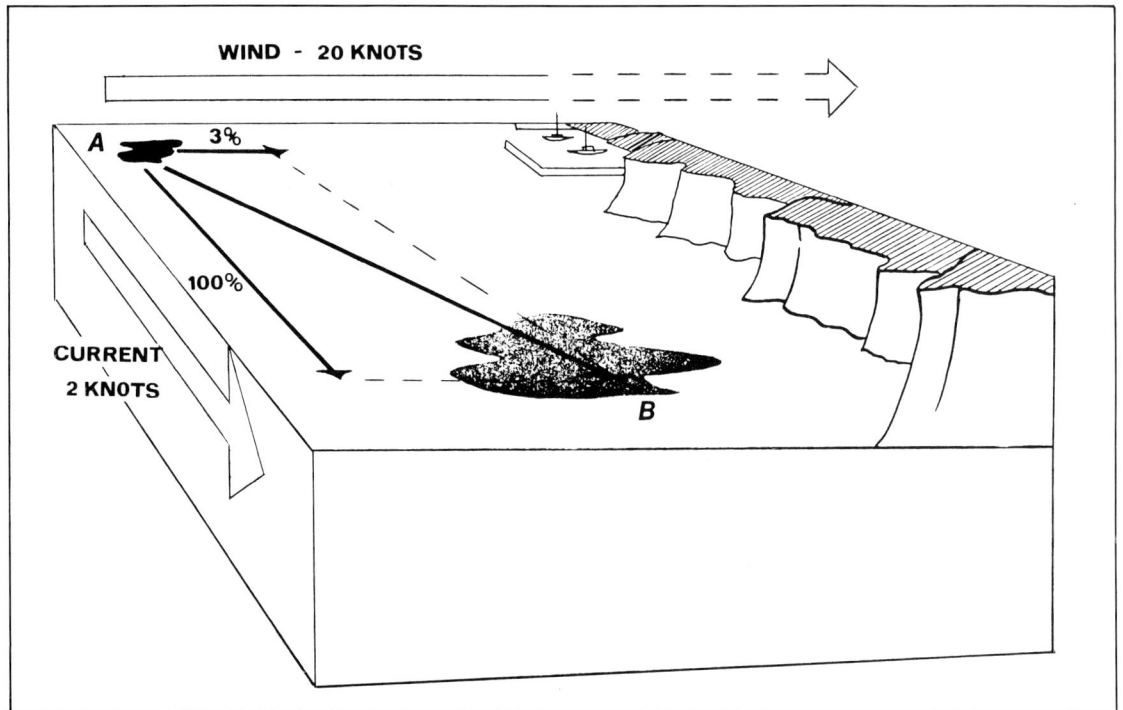

Figure 4:
The influence of 3% of the wind speed combined with 100% of the current speed results in the movement of oil from A to B.

AERIAL SURVEILLANCE AT SEA

However reliable an oil spill model may be, predictions of the fate and movement of oil slicks at sea should be verified through regular surveillance of the oil. This should be conducted from the air since observation from a vessel is highly inefficient.

Choice of Aircraft

Helicopters

Fixed-wing

Safety

The aircraft chosen for aerial observation must allow good all round vision and carry suitable navigational aids. Over nearshore waters the manoeuvrability of helicopters may be an advantage, for instance, in surveying an intricate coastline with cliffs, coves and islands. However, over the open sea there is less need for rapid changes in flying speed, direction and altitude and instead the speed and range of fixed-wing aircraft are generally desirable. For extensive surveys over remote sea areas, the extra margin of safety afforded by a twin or multi-engined aircraft is essential and may in any case be required by government regulations. Attention to safety must always be of paramount importance and the pilot will need to be consulted on all relevant aspects of a reconnaissance flight.

Flight plan

A flight plan should be prepared in advance using a chart of appropriate scale and taking account of any available information which may reduce the search area as much as possible. It is often advisable to draw a grid on the chart so that any position can be positively identified by a grid reference based, for example, on blocks of one square mile.

Search Pattern

'Ladder search'

A 'ladder search' is frequently the most economical method of surveying a large sea area (Figure 5). Since floating oil has a tendency to become aligned in long narrow windrows parallel to the direction of the wind, a ladder search across the wind will increase the chances of oil detection.

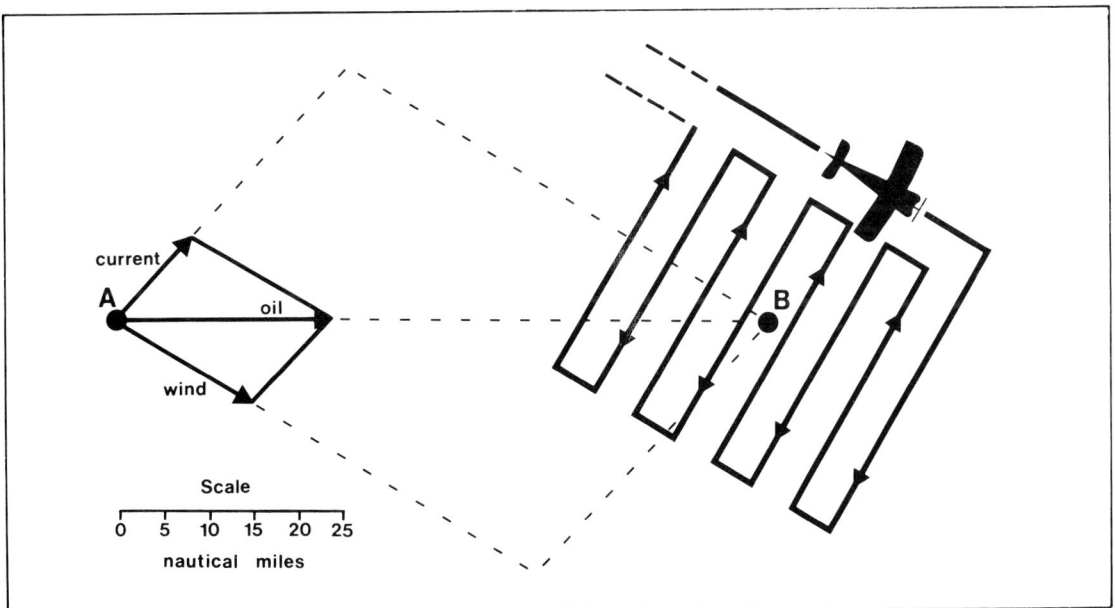

Figure 5: *Movement of oil from A to position B three days later, predicted by combining 100% of the current speed and 3% of the wind speed as shown. The arrows from A represent current, wind and oil movement for one day. A cross-wind ladder search pattern is shown over position B.*

Effect of sunlight

Search altitude

Haze and dazzle off the sea often affects visibility and the position of the sun may dictate the best direction to fly a search pattern. Sun glasses can give some relief from eye strain caused by strong light. Polarising lenses can assist the detection of oil at sea under certain light conditions due to the differences in light reflected from oil and water. The search altitude is generally determined by the visibility. In clear weather 500 metres (1600 feet) frequently proves to be optimum for maximising the scanning area without losing detail.

Navigation

However, it is necessary to drop to half this height or lower in order to confirm any sightings of floating oil or to examine its appearance. Over the open sea, away from any obvious reference points, it is easy to become disorientated. Ideally an observer will be able to consult the aircraft instrumentation for speed, direction and position, but it is worth ensuring beforehand that the instruments can be read without difficulty. In the absence of such aids, an observer with a suitable chart can keep track of course changes and positions by communicating with the pilot using the aircraft intercom.

Appearance of Oil at Sea

Confusing effects

Tank washings

From the air it is very difficult to distinguish between oil from spills and a variety of other unrelated phenomena. These include cloud shadows; ripples on the sea surface; seaweed patches in shallow water; differences in the colour of two adjacent water masses; river sediments and sewage discharges. It is necessary therefore to verify initial sightings of suspected oil by overflying the area at a sufficiently low altitude to allow positive identification. The appearance of tank washings and bilge discharges as a single elongated slick usually distinguishes them from accidental spills.

Colour

Crude or fuel oils spilled at sea undergo marked changes in appearance due to weathering. Initially, the thicker parts will usually appear as dense, black areas but as emulsification takes place the colour will change to brown, orange or yellow. In contrast, the thinner parts will have the appearance of irridescent or silver films.

Remote Sensing

Sensor types

A number of airborne sensors have been evaluated to assist the detection, mapping and quantification of oil on water, some of which can be used in conditions of poor visibility and at night. No one sensor is capable of providing sufficient information under all conditions and it is therefore necessary to combine a number of different devices to provide an operational system. The most commonly employed combination of sensors (Figure 6) are ultra-violet and infra-red line scanners (UVLS and IRLS) and side-looking airborne radar (SLAR).

SLAR

Side-looking airborne radar is particularly useful for obtaining information on the overall extent of an oil slick although it cannot give any indication of slick thickness. It relies on the calming effect created by oil on the sea. The device emits radiation in the microwave region and detects the differences in the echo signal from ordinary sea waves and oil-damped waves. Consequently the sensors are ineffective in calm sea conditions and, furthermore, other wave-damping phenomena such as wind and tide interactions may give similar signals to oil. However, SLAR is able to detect oil over a wide area (up to 20 miles either side of the aircraft when equipped with twin antennae) and can operate day and night. In addition, a particular advantage of radar systems is that microwave radiation is largely unaffected by fog and cloud.

IRLS

IRLS has a much smaller field of view than SLAR and although the sensors can operate at night, they cannot 'see' through fog and cloud. IRLS devices measure the natural radiation emitted from the sea in the thermal infra-red region and detect temperature differences at the sea surface. As a result, IRLS is not specific to oil and its presence has to be confirmed either visually or in combination with UVLS. However, IRLS can provide a broad indication of slick thickness in daylight due to differences in solar absorption.

UVLS

UVLS systems detect differences in the UV light reflected from the sea surface and so can only operate in daylight. Very thin films of oil can be detected with these sensors including those of biological origin.

An instrument package combining all three sensors would typically employ firstly SLAR to provide a rapid sweep over a wide area to give an indication of the presence of oil. This would then be confirmed by a more detailed search using UVLS and IRLS, with IRLS to provide qualitative information on slick thickness.

Signals from the sensors, together with details of time and position, are displayed on a monitor in the aircraft and with some systems can be relayed to vessels and ground stations. The data is usually stored on video tape and sections of interest can be printed on paper for a permanent record.

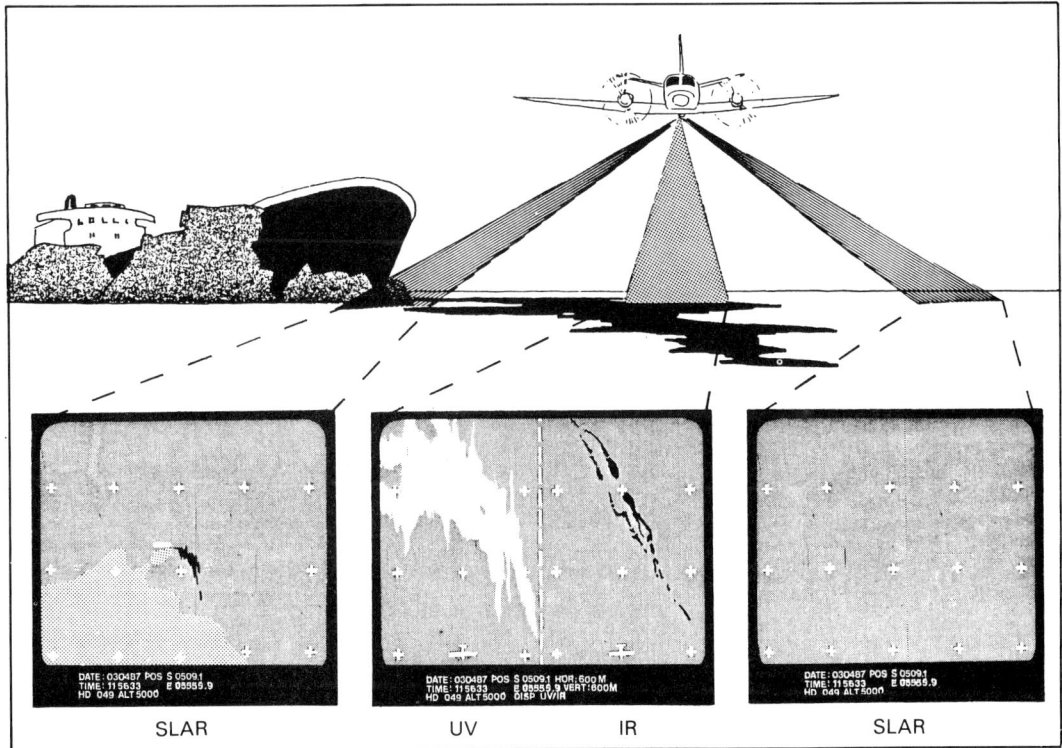

Figure 6:
Remote sensing aircraft fitted with Side-Looking Airborne Radar, Ultra-Violet Line Scan and Infra-Red Line Scan oil detection systems.

Visual Quantification of Floating Oil

An accurate assessment of the quantity of floating oil is virtually impossible due to the difficulty of gauging its thickness. At best, the correct order of magnitude can be estimated by considering certain factors. Oil spreads rapidly and most liquid oils will soon reach an average thickness of about 0.1 mm, characterised by a black or dark brown appearance. Similarly, the colour of sheen roughly indicates its thickness (see Table 3).

Appearance versus thickness

Cold water effects

A reliable estimate of water content in a 'mousse' is not possible without laboratory analysis, but accepting that figures of 50% to 80% are typical, approximate calculations of oil quantities can be made, given that most typical floating 'mousses' are 1 mm or more thick. However, it should be emphasized that the thickness of 'mousse' and other viscous oils is particularly difficult to gauge because of their limited spreading. Indeed in cold waters some oils with high pour points will solidify into unpredictable shapes and the appearance of the floating portions will belie the total volume of oil present. The presence of ice floes and snow in such conditions will add further confusion.

Table 3. A Guide to the Relation between Appearance, Thickness and Volume of Floating Oil

Oil Type	Appearance	Approximate thickness (mm)	Approximate Volume (m^3/km^2)
Oil sheen	silvery	0.0001	0.1
Oil sheen	irridescent	0.0003	0.3
Crude and fuel oil	black/dark brown	0.1	100
Water-in-oil emulsions ('mousse')	brown/orange	>1	>1000

Surface area

Percentage cover

In order to estimate the amount of floating oil it is necessary not only to gauge thickness, but also to determine the percentage area of the sea surface covered by oil, water-in-oil emulsion and sheen. Again, accurate estimates are complicated by the patchy incidence of floating oil. To avoid distorted views, it is necessary to look vertically down on the oil when assessing its distribution. By estimating the percentage coverage of each form of oil, the area covered relative to the total sea area affected can be calculated from timed overflights at constant speed or from position fixing equipment.

To illustrate further the process of estimating oil quantities, the following example is given and shown in Figure 7:

Example

During aerial reconnaissance flown at a constant speed of 150 knots, crude oil 'mousse' and silver sheen were observed floating within a sea area, the length and breadth of which required respectively 65 seconds and 35 seconds to overfly. The percentage cover of 'mousse' patches within the contaminated sea area was estimated at 10% and the percentage cover of sheen at 90%.

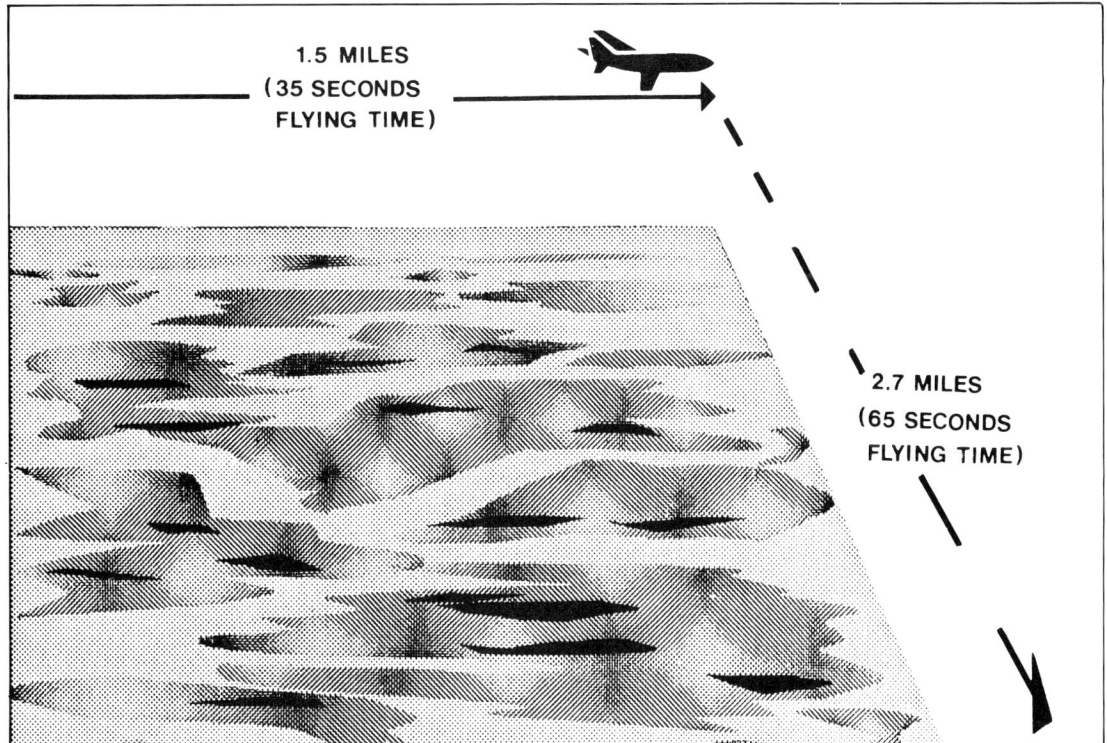

1.5 MILES
(35 SECONDS FLYING TIME)

2.7 MILES
(65 SECONDS FLYING TIME)

From this information it can be calculated that the length of the contaminated area of sea measured is:

$$\frac{65 \text{ (seconds) x } 150 \text{ (knots)}}{3600 \text{ (seconds in one hour)}} = 2.7 \text{ nautical miles}$$

Similarly, the width of the sea area measured is:

$$\frac{35 \times 150}{3600} = 1.5 \text{ n.m.}$$

This gives a total area of approximately 4 square nautical miles or 14 square kilometres. The volume of 'mousse' can be calculated as 10% (percentage coverage) of 14 (square kilometres) x 1000 (approximate volume in m^3 per km^2 — from Table 3). In this case it was estimated that 50% of the mousse was water, giving a volume of oil present as approximately 700m^3. A similar calculation for the volume of sheen yields 90% of 14 x 0.1 which is equivalent to approximately 1.3m^3 of oil.

This example also serves to demonstrate that although sheen may cover a relatively large area of the sea surface, it makes a negligible contribution to the volume of oil present. Hence, it is crucial to distinguish between sheen, thicker oil and emulsion.

RECOGNITION AND QUANTIFICATION OF OIL ON SHORELINES

Natural collecting points

A knowledge of the locations where floating debris collects is useful when predicting where oil may accumulate naturally on a coastline. Small coves and inlets as well as under jetties, piers and other man-made structures are examples of locations where oil can become trapped.

Shoreline type

Tidal shores

Photographic records

The appearance of stranded oil depends to a large extent on the type of coastline, which can vary from exposed rocky shores, through pebble and sand beaches to sheltered muddy wetlands. Oil pollution is seldom uniform in either thickness or coverage, unless the contamination is very heavy. Winds, waves and currents cause oil to be deposited ashore in streaks or patches rather than as a continuous cover. On tidal shores the affected zone can be comparatively wide, particularly on flat, sheltered beaches, but elsewhere the pollution may be confined to a narrow band close to the high water mark. When reporting shore pollution it should be specified which parts of the shore are covered by oil. Photographs are a very useful support to any description of the location and appearance of oil on shorelines. They also serve as a record against which later changes in the pollution may be compared.

Quantifying Stranded Oil

Visual inspection

When organising shore clean-up and monitoring its progress, an assessment should be made of roughly how much oil is present on a stretch of coastline. The variable distribution of the oil can cause serious errors unless the task of estimating the quantity is approached with care. The assessment is largely a visual one and will be impossible if the oil is effectively hidden from view for example, by sand, snow or vegetation such as mangroves. Where the oil is visible, the problem can be met in two stages.

Aerial reconnaissance

Verification

First, the overall extent of the contamination along a coastline can be estimated and marked on a large-scale chart or map. In the case of a major spill, a helicopter over-flight is usually the most efficient and convenient way of gaining a general impression. A fixed-wing aircraft generally travels too fast for a good coastal inspection at low altitude. Aerial surveillance should always be combined with spot checks on foot since many shoreline features, such as seaweed and black sand look like oil when viewed from a distance. Careful attention should be given to identifying locations where the character of the shoreline changes or where the oil coverage appears to change. Binoculars are useful, but any changes detected at a distance should always be verified by a closer inspection.

Quantifying representative areas

The second stage of quantifying stranded oil involves selected representative areas of shoreline for a calculation of the amount of oil present. The area chosen should be small enough to allow an accurate estimate of oil volume in a reasonable time, yet large enough to be representative of the whole shore section similarly affected. The exercise has to be repeated on other sections where the degree of oil coverage may be different.

Source of errors

Photographs

Quantifying stranded oil in this way only yields an approximate figure due to several inescapable sources of error. On a sandy beach oil may soak into the beach substrate. If the saturation is uniform, a quarter of the depth of oily sand may be taken as representing oil. The presence of debris or stones and crevices on rocky shores can be an added complication, and when calculating oil volumes the occurrence of water-in-oil emulsions can be misleading. If in some situations it proves impracticable to use the relatively time-consuming methods outlined above, it should always be possible to describe the degree of pollution as either light, moderate or heavy by comparing the oiled shoreline with reference photographs (Figure 8).

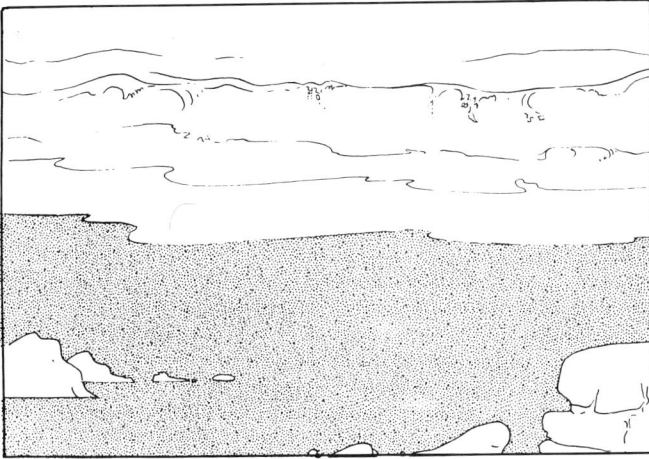

Figure 8a:
Unpolluted sandy beach.

Figure 8b:
Light oil pollution —
Less than 10 ml oil/m².

Figure 8c:
Moderate oil pollution —
10 ml-1 litre oil/m².

Figure 8d:
Heavy oil pollution —
1-100 litres oil/m².

Sampling Procedures

Source of pollution

Analysis

Hindcasting

Oil pollution may lead to claims for compensation. Evidence will be required linking the damage or costs incurred to the alleged polluter. Sometimes the link is easy to establish, but on occasions chemical analysis of oil taken from the suspected source and the polluted area is necessary. It is not always possible for a potential claimant to secure a reference sample of oil from a suspected source, but analysis of a pollution sample alone can occasionally yield sufficient circumstantial evidence to indicate the probable source. With knowledge of oil weathering processes together with recorded winds and currents it is possible to deduce by 'hindcasting' the approximate position where a pollutant may have been released.

Legal requirements

Subsamples

Whatever the particular circumstances of an incident, it is important that careful attention is paid to oil sampling and sample storage procedures so as to support the credibility of any analytical evidence. Legal advice may be necessary to ensure that the actions taken during sampling conform with legal requirements. Replicate samples should be taken, preferably by splitting the main sample into subsamples in the presence of an independent witness. Although for most analytical purposes a few millilitres of oil are sufficient, the effort spent in obtaining larger volumes (250 ml) will reduce the expense of handling and processing minimal quantities. Usually a minimum of three subsamples are necessary to ensure that the main interested parties have the option of independent analysis. Chemical analysis is relatively costly and it is usually only necessary to analyse key samples to settle a dispute.

Labelling and storage

To avoid contamination of samples during storage, glass jars are recommended and these should be sealed and securely labelled. The information carried on the label should include sampling location, date, time, name, signature and address of sample collector and witness. Precautions must be taken to prevent deterioration of stored samples by evaporation and biodegradation. Refrigeration of sealed samples in the dark at less than 5°C is recommended.

EFFECTS OF OIL SPILLS

Oil spills can have a serious economic impact on coastal activities and on those who exploit the resources of the sea. In most cases, such damage is temporary and is caused primarily by the physical properties of oil creating nuisance and hazardous conditions. The impact on marine life is compounded by toxicity and tainting effects resulting from the chemical composition of oil, as well as by the diversity and variability of biological systems and their sensitivity to oil pollution.

The extent of the damage caused by a spill does not always reflect the quantity of oil spilled. A little oil in a sensitive area can do considerably more harm than a large quantity on a desolate rocky shore.

Impact of Oil on Coastal Activities

Recreational Activities

Visual impact

Contamination of coastal amenity areas is a common feature of many oils spills leading to public disquiet and interference with recreational activities such as bathing, boating, angling and diving. Hotel and restaurant owners, and others who gain their livelihood from the tourist trade can also be affected. Because of their visual impact, persistent oils and their residues cause the most nuisance and concern, with the greatest effect likely to be just before or during the main tourist season. The disturbance to coastal areas and to recreational pursuits from a single oil spill is comparatively short-lived and any effect on tourism is largely a question of restoring public confidence once clean-up is completed.

Industry

Power stations

Desalination plants

Industries that rely on a continuous supply of clean seawater for their normal operations can be adversely affected by oil spills. Power stations, in particular, are often located close to the coast in order to have access to the very large quantities of water required for cooling purposes (Figure 9). If substantial quantities of oil are drawn through intakes, contamination of the condenser tubes may result, requiring a reduction in output or total shutdown whilst cleaning is carried out. Similarly, the normal operation of desalination plants may be disrupted by oil, causing water supply problems for consumers.

Figure 9:
Industrial plant with water intakes protected by a floating boom across the intake channel.

Shipyards

Ports

Shipyards with slipways and drydocks for construction and repair work may be affected by oil spills causing damage to unpainted or newly painted surfaces and creating hazardous working conditions. Interference with shipping may result from oil spills and clean-up operations, particularly when they take place in harbours and port approaches. The installation of booms or closure of lock gates to contain oil may cause delays. Direct contamination of jetties as well as mooring lines and ships' hulls is a common occurrence.

Fire hazard

Other routine harbour activities such as ferry services and lock operations can be disrupted, particularly after a spill of light crude oil, gasolene or other flammable material. Welding and the use of spark-generating machinery may have to be suspended as long as a fire hazard persists. In this way even small spills in a busy port can have considerable repercussions.

Biological Effects of Oil

The effects of oil on marine life can be considered as being caused by either its physical nature (physical contamination and smothering) or by the chemical components of the oil (toxic effects and accumulation leading to tainting). Marine life may also be affected by clean-up operations or indirectly through physical damage to the habitats in which they live.

Natural changes

Populations of plants and animals in the sea are subject to considerable natural fluctuations in numbers brought about, for example, by changes in climatic and hydrographic conditions and the availability of food. Thus the species composition and age structure of the various populations within a particular marine habitat are far from constant but instead are in a state of dynamic balance. In view of this it is usually extremely difficult to assess the effects of an oil spill and to distinguish changes caused by the oil from those due to natural variability.

Natural losses

The different life stages of a species may show widely different tolerances and reactions to oil pollution. Usually the eggs, larval and juvenile stages will be more susceptible than the adults. However, many marine species produce very large numbers of eggs and larval stages to overcome natural losses. This will normally result in less than 1 in 100,000 eggs or larvae surviving to maturity but the excess production provides a reservoir to compensate for any extreme losses due to adverse local conditions. These facts make it unlikely that any localised losses of eggs or larvae caused by an oil spill will have a discernible effect on the size or health of future adult populations.

Recovery

The ability of animal and plant populations to recover from an oil spill and the time taken for a normal balance in the habitat to be re-established depends upon the severity and duration of the disturbance and the recovery potential of the individual species. Abundant organisms with highly mobile young stages produced regularly in large numbers may repopulate an area rapidly when pre-spill conditions are restored, whereas populations of long-lived, slowly maturing species with low reproductive rates may take many years to recover their numbers and age structure. In general, the rate of recovery in tropical regions is much faster than in cold environments.

Restoration

Whilst it may be possible to restore the physical characteristics of an oiled habitat to near its pre-spill condition, the extent to which its biological recovery can be enhanced is severely limited. Although the cleaning of mangroves and salt marshes, and replanting with seedlings, may be feasible in some situations, care needs to be exercised to ensure that the area is not physically damaged since this may be more destructive in the longer term than the loss of the vegetation. The replacement of animals is virtually impossible and although some species can be bred and released or be moved from undamaged areas (e.g. certain birds, mammals, reptiles and fish) it is highly improbable that such programmes will accelerate the natural recovery of a complex marine habitat.

Impact of Oil on Specific Marine Habitats

The following summarises the impact that oil spills can have on selected marine habitats. Within each habitat a wide range of environmental conditions prevail and often there is no clear division between one habitat and another.

Open Waters and Seabed

Plankton

Plankton is a term applied to floating plants and animals carried passively by water currents in the upper layers of the sea (Figure 10). They form the base of the marine food web and include the eggs and young stages of fish, shellfish and many bottom-living animals. Their sensitivity to oil pollution has been demonstrated experimentally. In the open sea, the rapid dilution of naturally dispersed oil and its soluble components, as well as the high natural mortality and patchy, irregular distribution of plankton, make significant effects unlikely.

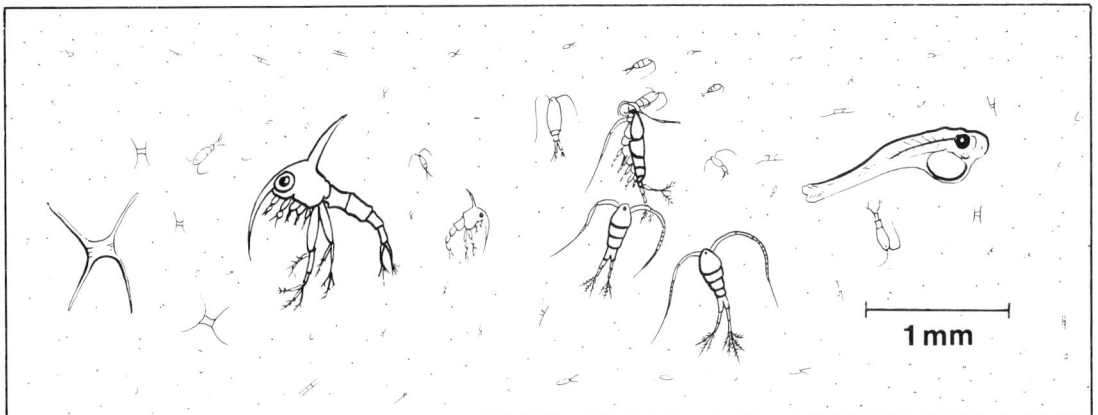

Figure 10:
Plankton.

Fish and mammals

Large swimming animals such as squid, fish, turtles, whales and dolphins are highly mobile and rarely affected in offshore waters even in major oil spills. In coastal areas some marine mammals, such as seals, and reptiles, such as turtles, may be particularly vulnerable to adverse effects from oil contamination because of their need to surface to breathe and to leave the water to breed. Adult fish living in nearshore waters and juveniles in shallow water nursery grounds may also be at risk from exposure to dispersed or dissolved oil.

Benthos

Toxicity

Sediments

Plants and animals living on the sea bed (benthos) also form an important part of the food web and in nearshore waters many of the animals ('shellfish') and some seaweeds, such as kelp, are exploited commercially. The risk of surface oil slicks affecting the sea bed in offshore waters is minimal, but in shallow waters oil droplets may reach the bottom, particularly during periods of rough weather. Fresh crude oils and light refined products with a high proportion of toxic components can cause local damage to seagrass beds, and to various animals such as clams, sea urchins and worms. The incorporation of oil into sediments can lead to residence times of several years in localised areas, with the possibility of sub-lethal effects and tainting of commercial species. Weathered oil may accumulate sediment particles and sink, especially after temporary stranding, possibly causing damage to benthic species.

Shorelines

Intertidal organisms

Shorelines, more than any other part of the marine environment, are exposed to the effects of floating oil. The impact may be particularly great where large areas of rock, sand and mud are uncovered at low tide. Whilst intertidal animals and plants are able to withstand short-term exposure to adverse conditions, they may be killed by toxic oil components or physically smothered by viscous and weathered oils and emulsions. Animals may also become narcotised by the oil such that they become detached from rock surfaces or emerge from burrows. They are then susceptible to predators or to being washed into an area where they cannot survive. Recolonisation of a shoreline by the dominant plant and animal species can be rapid: on rocks the initial stage is usually the settlement of seaweeds followed by the slower return of grazing animals. However, the complete re-establishment of a normal balance may, in extreme situations, take many years.

Recoloni-sation

Wetlands

Salt marshes

Salt marshes occurring in sheltered waters in temperate and cold regions are characterised by dense low vegetation on mud flats drained by a network of channels (Figure 11). The organic input from the marsh provides the basic source of food for a rich and diverse fauna of worms, snails, clams and crabs which in turn are eaten by birds congregating in large numbers at low tide, especially at certain times of the year.

Figure 11:
Salt marsh.

Effect on vegetation

Marsh vegetation shows greater sensitivity to fresh light crude or light refined products than to weathered oils which cause relatively little damage. Oiling of the lower portion of plants and their root systems can be lethal whereas even a severe coating on leaves may be of little consequence, especially if it is outside the growing season. More widespread damage can be expected from repeated contamination or if oil penetrates into sediments where it may persist for several years. Similarly, if oil reaches the inner portion of marshes during a period of extreme high tides, the residence time may be prolonged, affecting the plants as well as birds that feed and roost there.

Mangroves

In tropical regions, mangrove forests are widely distributed and replace salt marshes on sheltered coasts and in estuaries (Figure 12). Mangrove trees have complex breathing roots above the surface of the organically rich and oxygen-depleted muds in which they live. The network of roots and trapped sediment create productive habitats for fish, shrimps, crabs, oysters, snails, mussels and other animals living directly or indirectly on the nutrients from fallen mangrove leaves. Mangrove forests also provide food and shelter for the young stages of commercially important fish and prawns. Fishing in the creeks and drainage channels and collection of shellfish from among the prop roots sustain village communities often living at subsistence level. The wood is also used for burning, building and tanning.

Commercial importance

Oil may block the openings of the air breathing roots of mangrove trees or interfere with their salt balance, causing leaves to drop and the trees to die. The root systems can be damaged by fresh oil entering nearby animal burrows and the effect may persist for some time inhibiting recolonisation by mangrove seedlings. The long-term effects on the associated animals are likely to be less severe.

Figure 12:
Tropical marine habitats: mangroves, seagrass beds and coral reef.

Corals

Sensitivity of coral reefs

Reef building corals are found off most tropical coastlines and islands, in shallow warm waters of suitable salinity and clarity (Figure 12). The living coral grows on the calcified remains of dead coral colonies which form overhangs, crevices and other irregularities inhabited by a rich variety of fish and other animals. To the land side of the reef crest and reef flat, a lagoon is often found which is a low energy environment usually with a sand bottom and seagrass beds, protected by the outer reef. If the living coral is destroyed, the reef itself may be subject to wave erosion.

Toxic effects

Coral reefs are generally submerged and it is only if they are briefly exposed to air at low tide that they are vulnerable to physical coating by floating oil. Because of the turbulence and wave action characteristic of reefs, the corals may be exposed to naturally dispersed oil droplets. The effects of oil on corals and their associated fauna are largely determined by the proportion of toxic components, the duration of oil exposure as well as the degree of other stresses. Observations of oiled corals suggest that several sub-lethal effects may occur such as interference with reproductive processes, abnormal behaviour and reduced or suspended growth. Most of the effects are temporary but similar effects on the associated reef fauna can have greater repercussions since narcotised animals may be swept away from the protection of the reef by waves and currents.

Sub-lethal effects

Birds

Susceptible types

Birds which congregate in large numbers on the sea or shorelines to breed, feed or moult are particularly vulnerable to oil pollution (Figure 13). They include auks, penguins and ducks, but other more solitary species such as pelicans, cormorants and gannets which dive to feed can also be affected. Although oil ingested by birds during preening may be lethal, the most common cause of death is from drowning, starvation and loss of body heat following damage to the plumage by oil. Feathers and down matted with oil lose their waterproofing and insulating properties.

Figure 13:
Sea bird colony.

Population effects

There are many examples of oil spills having caused large bird mortalities at sea leading to concern for the survival of populations. Many seabirds have long lives, delayed maturity and low rates of reproduction and these factors taken together have been thought to hinder rapid recovery. However, recent research indicates that not all mature birds breed at any given time and that these individuals form a reservoir for replacement of killed birds. Oil spill mortalities may not therefore have a detectable impact on the breeding populations of sea birds except possibly in the case of isolated colonies with limited potential for recolonisation from elsewhere and those exposed to other stresses, for example, through being at the limit of their geographical range.

Cleaning and rehabilitation

The cleaning and rehabilitation of oiled birds requires trained personnel and inevitably causes the birds considerable distress. While individual birds may be saved, the rationale for cleaning oiled birds is usually based more on animal welfare than any expectation of promoting the recovery of bird populations, except in the case of endangered species.

Impact of Oil on Fisheries and Mariculture

Fishing boats and gear

An oil spill can directly damage the boats and gear used for catching or cultivating marine species. Floating equipment and fixed traps extending above the sea surface are more likely to become contaminated by floating oil whereas submerged nets, pots, lines and bottom trawls are usually well protected, provided they are not lifted through an oily sea surface. However, they may sometimes be affected by sunken oil.

Effects on catches

Reduced catches of fish, shellfish and other marine organisms are occasionally reported after an oil spill. Most often this is due to a reduction in fishing effort although on rare occasions mortalities can be caused by physical contamination or close contact with freshly spilled oil in shallow waters with poor water exchange. It is sometimes suggested that fish and shellfish stocks will be depleted for a number of years after a spill as a result of damage to eggs and larvae. However, experience from major oil spills has shown that the possibility of such long-term effects is remote because the normal over-production of eggs provides a reservoir to compensate for any localised losses.

Fish farms

Seaweed

Cultivated stocks, such as caged fish, are more at risk from an oil spill: natural avoidance mechanisms may be prevented, and the oiling of cultivation equipment may provide a source for prolonged input of oil components and contamination of the organisms (Figure 14a). Cultured seaweed is particularly vulnerable in tidal areas where it may become contaminated with oil at low tide. Seaweeds such as kelp, suspended in deep water from floating structures, are better protected (Figure 14b).

Figure 14a:
Mussel cultivation.

Figure 14b:
Seaweed cultivation.

Quantifying damage

It is comparatively easy to determine oil spill mortalities in a cultivated stock of known size. Losses can be quantified by comparing post-spill production with yields and market values in previous years or in adjacent unaffected areas. The situation in the case of naturally occurring species is frequently far more difficult since accurate stock assessment is impossible and any dead individuals are likely to be consumed by scavengers. Catch statistics are rarely sufficiently detailed to enable any decline due to an oil spill to be isolated from changes brought about by other factors such as variable fishing effort and natural fluctuations in the size of the stock.

Market confidence

Tainting

An oil spill can cause loss of market confidence since the public may be unwilling to purchase marine products from the region irrespective of whether the seafood is actually tainted. Because of the serious economic consequences arising from a loss of sales, considerable care is frequently exercised to prevent contaminated fish and shellfish from reaching the market. Ideally, this should involve organised tasting by qualified personnel at the time of the spill. To eliminate the effect of subjective assessments, the tasters must correctly establish the absence and presence of taint in clean and deliberately oiled controls respectively. These control samples should be introduced amongst the test samples in a random manner so as to give the taster no clues. A number of replicate tests should be made by each taster to ensure that the results are statistically significant.

Fishing bans

Bans on the fishing and harvesting of marine products may be imposed following a spill, both to maintain market confidence and to protect fishing gear and catches from contamination. The area covered by a ban should be related directly to the location and extent of oil slicks and the justification for its maintenance reviewed frequently in the light of the position of floating oil and the results of taste tests. Such bans should not be kept in force beyond the relatively short time normally necessary while oil is present. Rearing of cultured species often follows a strict timetable and delays designed to protect the culture from oil can prove counter-productive if growth is impaired and the subsequent harvest is poor as a result.

CONTINGENCY PLANNING

Tankers

Plans covering areas where a wide range of oil types are handled or where tankers pass in transit, cannot anticipate the impact of a spill. It is therefore important that the type of oil spilled is established at the earliest opportunity so that its fate can be predicted and the appropriate clean-up techniques employed.

Fixed installations

For oil terminals and offshore oil fields, where a limited number of oil types are involved, an appreciation of the likely fate of potential spills is valuable when drawing up contingency plans. Information on the prevailing winds and currents throughout the year will indicate the resources where oil spill impact is most likely. Data on the types of oil handled can enable predictions to be made regarding the lifetime of slicks and the quantity and nature of the residue which may require a clean-up response. It will also assist in the selection of appropriate clean-up equipment to be held in readiness for spills.

Priorities for protection

Sensitivity maps

Because of the difficult decisions that will be required during an oil spill in order to mitigate damage and to resolve conflicts of interest, much can be done at the contingency planning stage to identify sensitive areas and to determine priorities for protection. The mapping of sensitive areas can be a useful starting point. Detailed consideration should be given to the likely impact that a spill would have on each habitat or activity, taking into account any seasonal variability. Attention should then be given to identifying areas to be protected and their order of priority. This will never be easy since the value of each resource to the community will depend upon the weight given to environmental, recreational, economic and political considerations. This may require a wide range of data to be gathered and evaluated. If properly conducted, such studies of the resources at risk in an area can also form a basis for quantifying any damage caused by a spill.

Response decisions

Having determined priorities for protection, attention can be given to designating appropriate clean-up measures. It is necessary to make a realistic assessment of the feasibility of employing various techniques since a recommendation to avoid the more ecologically damaging response options may result in the adoption of ineffective techniques and greater damage to other habitats or activities.

POINTS TO REMEMBER

1. The total annual input of petroleum hydrocarbons into the sea is approximately 3 million tonnes of which some 15% is due to accidents related to exploration, production and transportation activities.

2. When oil is spilled at sea its fate is determined by a number of processes, some of which lead to its removal from the sea surface whilst others cause it to persist.

3. Whilst oil is eventually assimilated into the marine environment, the timescale depends upon the physical and chemical characteristics of the oil and on the weather and climatic conditions.

4. Simple calculations and computer models can be used to predict the movement and overall fate of an oil spill, but the results should be verified by regular aerial reconnaissance.

5. The changes that spilled oil undergo during weathering will affect the choice of appropriate clean-up techniques and the nature of the damage that will be caused.

6. Persistent oil may seriously affect the visual appeal and use of coastal amenity areas and can interfere with the normal working of power stations and other industrial plants which require a continuous supply of clean sea water.

7. Effects on marine life are caused by the physical nature of the oil (physical contamination and smothering) and by its chemical composition (toxic effects and tainting).

8. An oil spill can contaminate fishing equipment and mariculture facilities and cause loss of market confidence in marine products. Populations of adult fish are rarely, if ever, affected.

FURTHER READING

Burridge, J. and Kane, M. (Eds.) (1985) Rehabilitating oiled sea birds: A field manual. American Petroleum Institute, Washington D.C. 79 pp.

Butt, J.A., Duckworth, D.F., and Perry, S.G. (Eds.) (1986) Characterization of spilled oil samples — Purpose, sampling, analysis and interpretation. John Wiley & Sons, Chichester, England. 95 pp.

Clark, R.B. (Ed.) (1982) The long-term effects of oil pollution on marine populations, communities and ecosystems. The Royal Society, London. 259 pp.

CONCAWE (1983) Characteristics of petroleum and its behaviour at sea. CONCAWE, The Hague, Netherlands. Report No. 8/83. 47 pp.

Exxon (1985) Fate and effects of oil in the sea. Exxon Background Series. Exxon, New York. 12 pp.

Ferguson Wood, E.J. and Johannes, R.E. (Eds.) (1975) Tropical marine pollution. Elsevier Scientific Publishing Company, Amsterdam. 192 pp.

Fincham, A.A. (1984) Basic marine biology. Cambridge University Press, Cambridge, England. 144 pp.

GESAMP (1977) Impact of oil on the marine environment. FAO, Rome. 250 pp.

ITOPF (1981) Aerial observation of oil at sea. Technical Information Paper No. 1. ITOPF, London. 8 pp.

ITOPF (1983) Recognition of oil on shorelines. Technical Information Paper No. 6. ITOPF, London. 7 pp.

ITOPF (1985) Effects of marine oil spills. Technical Information Paper No. 10. ITOPF, London. 8 pp.

ITOPF (1986) Fate of marine oil spills. Technical Information Paper No. 11. ITOPF, London. 8 pp.

Jordan, R.E. and Payne, J.R. (1980) Fate and weathering of petroleum spills in the marine environment. Ann Arbor Science, Ann Arbor, U.S.A. 174 pp.

Massin, J.M. (Ed.) (1984) Remote sensing for the control of marine pollution. NATO Challenges of Modern Society, Volume 6. Plenum Press, New York. 466 pp.

National Research Council (1985) Oil in the sea. Inputs, fates and effects. National Academy Press, Washington D.C. 601 pp.

Neff, J.M. & Andersen, J.W. (1981) Response of marine animals to petroleum and specific petroleum hydrocarbons. Applied Science, London. 177 pp.

Nelson-Smith, A. (1972) Oil pollution and marine ecology. Paul Elek (Scientific Books) Ltd., London. 260 pp.

II CONTAINMENT AND RECOVERY

When oil is spilt on the sea surface, its removal is often desirable. The most common approach is to use some form of barrier to halt, or minimise, the spread of the oil and to concentrate it into a thick layer so that it can be recovered using a pump or skimmer.

This section examines the various techniques for containment and recovery with particular emphasis on the main design features and performance of booms and skimmers, their modes of operation and requirements for maintenance.

CONTENTS

CONTAINMENT

The containment of floating oil for subsequent recovery or its diversion away from sensitive areas calls for the use of some form of barrier. Many different types of oil barriers have been developed. These include commercially available floating booms, netting systems, sorbent booms, improvised booms and barriers, bubble barriers and chemical barriers. Selection of the most appropriate barrier will depend upon the particular conditions as well as availability. Since commercially available booms are the most common form of barrier used in oil spill control they are described in greatest detail in this section.

Commercially Available Booms

Design features

Designs vary considerably but all normally incorporate the following features:

1. freeboard to prevent or reduce splashover;
2. sub-surface portion (skirt) to prevent or reduce escape of oil under the boom;
3. flotation by air or some buoyant material;
4. longitudinal tension component (chain, wire or boom fabric itself) to withstand effects of winds, waves and currents.

Boom designs fall into two broad categories:

Types of booms

CURTAIN BOOMS provide a continuous sub-surface skirt or flexible screen supported by a solid or air flotation chamber usually of circular cross-section (Figure 1). Air flotation booms take up only a small storage area when deflated, whereas solid flotation booms, although more resistant to damage, are bulky in storage. Curtain booms generally have good wave-following capabilities, moderate escape velocities and are reasonably easy to clean.

FENCE BOOMS with a flatter cross-section are held vertically in the water by integral or external buoyancy (Figure 2). Solid flotation is most frequently used for fence booms but if external floats are used, turbulence may be generated leading to escape of oil at low water velocities. Such designs are bulky in storage and difficult to clean. In general, fence booms are more suitable for calmer waters where current velocities are low.

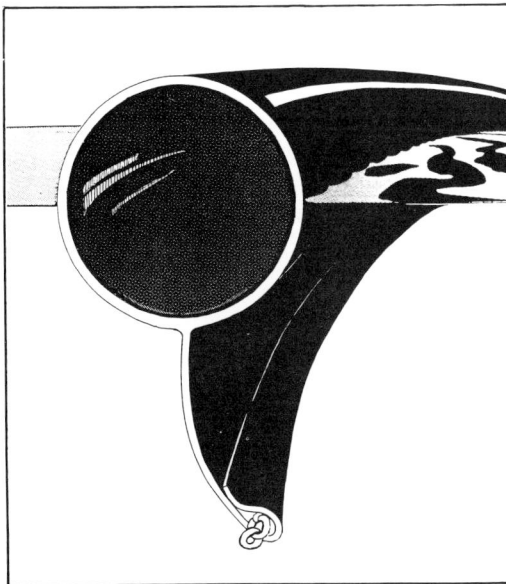

Figure 1:
Curtain boom with air flotation. Combined ballast and tension chain fitted in a pocket at the bottom of the skirt.

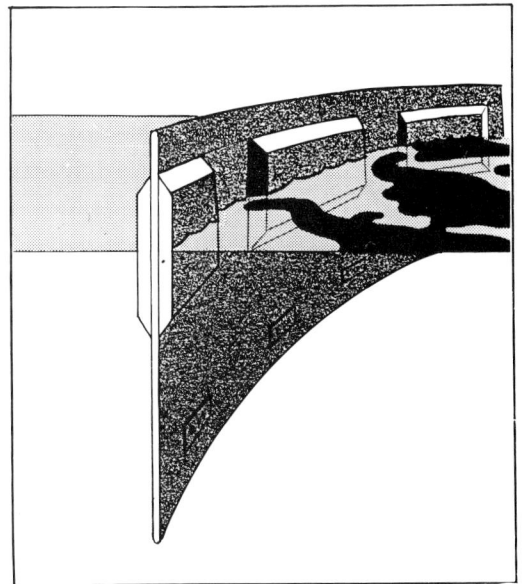

Figure 2:
Fence boom with solid flotation. Ballast weights fitted at intervals along the skirt.

Common features

Many curtain and fence booms have similar features including bracing struts and/or integral ballast to keep them upright in the water, connectors for joining sections together as well as towing and anchoring points.

Performance/Limitations

Currents

The most important characteristic of a boom is its oil containment or deflection capability, determined by its behaviour in relation to water movement. It should be flexible to conform to waves yet be sufficiently rigid to retain as much oil as possible. No boom can contain oil against water velocities much above 1 knot (0.5 metres per second) acting at right angles to it. The way in which oil escapes, and its relation with water velocity is as much a function of oil type as boom design. Low viscosity oils escape at lower velocities than more viscous materials. With the latter, the oil tends to accumulate at the boom face and to flow vertically down and under the skirt whereas low viscosity oils are carried under the boom as droplets sheared from the underside of the oil layer (Figure 3). Besides river and tidal currents, wind and waves can generate water velocities in excess of the escape velocity as well as causing splashover of contained oil. Oil escape can also result from turbulence along a boom and therefore a uniform profile without projections is desirable.

Wind, waves

Turbulence

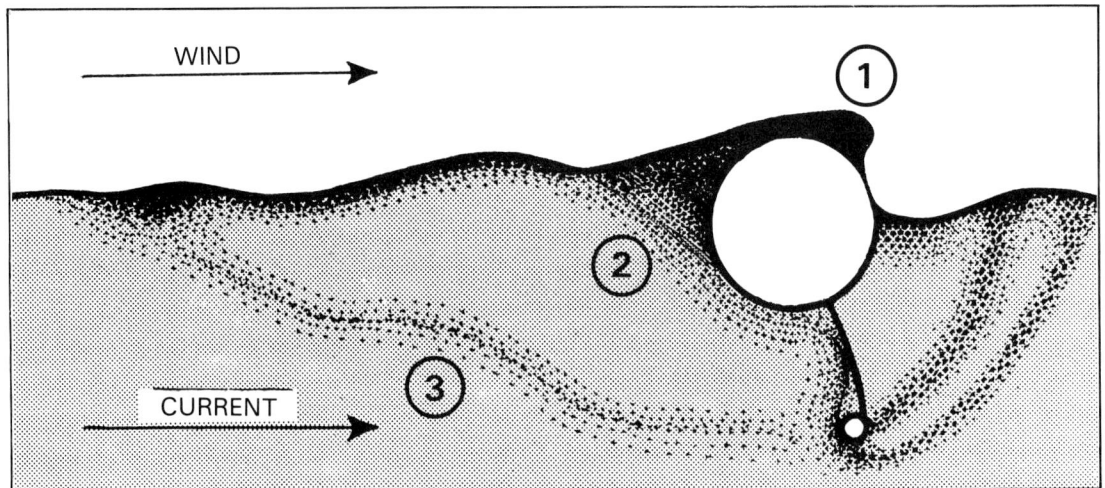

Figure 3:
Escape of oil from a boom
1) splashover by wave action
2) flow down the face of the boom
3) droplets sheared from the underside of the contained slick

Boom size

Connectors

The size and length of boom sections are also important considerations. The optimum size of a boom is largely related to the sea state in which it is to be used. As a general rule, the minimum freeboard to prevent oil splashover should be selected. The depth of skirt should be of similar dimensions to the freeboard. While short section lengths can make booms easier to handle and can protect the integrity of the boom as a whole should one section fail, these advantages must be weighed against the difficulty and time taken to connect sections effectively. Connections interrupt the boom profile and, wherever possible, should not coincide with the point of heaviest oil concentrations. The design of connectors should allow easy fastening and unfastening during deployment and whilst the boom is in the water.

Strength

Ease of deployment

Other important characteristics are strength, ease and speed of deployment, reliability, weight and cost. A boom must be sufficiently robust for its intended purpose and it must tolerate inexpert handling, since trained personnel are not always available. Structural strength and durability are required particularly to withstand the forces of water and wind on a boom when it is either towed or moored. Ease and speed of deployment combined with reliability are clearly very important in a rapidly changing situation and may strongly influence the choice made.

Forces Exerted on Booms

Currents

To estimate the approximate force F_C (kg) exerted on a boom with a sub-surface area A_S(m²) by a current with velocity V_C (knots) the following formula can be used:

$$F_C = 26 \times A_S \times V_C^2$$

Thus, the force acting on a 100 m length of boom with a 0.6 m skirt placed at right angles to a 0.5 knot water flow would be

$$F_C = 26 \times (0.6 \times 100) \times (0.5)^2 = 390 \text{ kg (force)}$$

When a boom is towed, its velocity through the water should be entered as V_C in the above formula. It can be seen that doubling the current velocity or towing speed would entail a four-fold increase in load.

Windage

The force (F_W) exerted by wind (V_W) directly on the freeboard (A_f) of the boom, can also be considerable. A similar formula can be used to estimate windage:

$$F_W = 26 \times A_f \times \left(\frac{V_W}{40}\right)^2$$

For example, the force on a 100 metre length of boom with a 0.5 metre freeboard in a 15 knot wind would be:

$$F_W = 26 \times (0.5 \times 100) \times \left(\frac{15}{40}\right)^2 = 183 \text{ kg force}$$

Combined forces

In the above example the combined forces of current and wind would be 573 kg if they were acting in the same direction on a rigid barrier. In fact booms are flexible and form a curve. In addition, the boom would be towed or moored at an angle to the flow. Both these factors lead to a reduction of the forces acting on the boom so that a considerable safety margin is included in the result of this calculation. Nevertheless, it provides a useful guide to the magnitude of such forces and can help in the selection of appropriate moorings or towing vessels.

Netting Systems

Advantages

The use of nets to recover solid tar balls is an obvious application and the extension of their use to contain viscous oils theoretically presents a number of advantages over the use of conventional booms. In particular, the open structure should offer less resistance to water movement so that light but strong sections could be manufactured which might realistically be long enough to enclose oil scattered over a wide area of sea. As a result of the lower resistance of nets to movement through the water, it should also be possible to operate in faster currents or to sweep or trawl the sea surface at higher speeds than can be achieved with conventional booms.

Designs

Two basic designs of net have so far been developed which draw on experience from the fishing industry (Figures 4a and 4b, overleaf): a long double net based on the purse seine method of fishing which can be used to corral or collect floating oil or which can be moored to protect sensitive areas; and a trawl net with a detachable 'cod-end' which can be towed along the sea surface.

Experience

Although neither design has yet been fully evaluated during an actual oil spill, large scale field trials show some promise, especially in the case of the purse seine type when used to corral and retain floating oil. However, once oil has been adsorbed onto the net the mesh becomes blocked and the oil retention capabilities are similar to conventional booms.

Figure 4a:
Netting system of the purse seine type for oil containment and recovery using two vessels to corral floating oil.

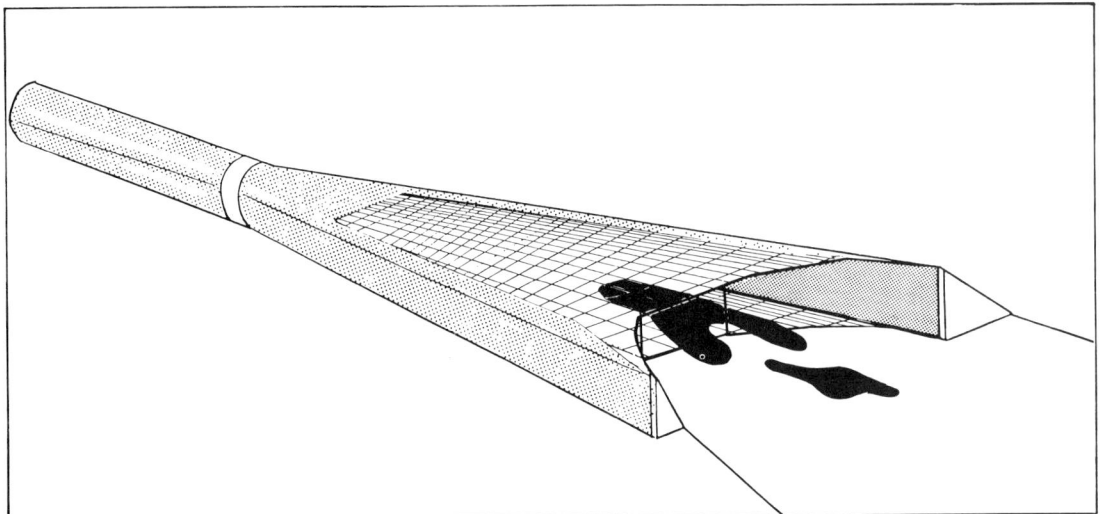

Figure 4b:
Oil trawl for collecting floating solid oil into a detachable cod-end.

Sorbent Booms

*Construc-
tion*

Uses

Sorbent booms usually consist of a tube of netting or some other fabric filled with a synthetic or natural sorbent material. Booms constructed of sorbent material have little inherent strength and, in some applications, may require additional support. Some also need extra flotation to prevent them sinking when they become saturated with oil and water. They are normally only used in areas of low current velocity to collect thin films of oil, since their recovery efficiency decreases rapidly once the outer layers of the sorbent material become saturated with oil. The handling and disposal of oil-soaked sorbent booms can also cause considerable problems. The use of sorbents is further discussed in the section on Recovery.

Improvised Booms and Barriers

Construc-
tion of
floating
booms

When purpose-built equipment is not available, it is possible to improvise successfully using booms made with locally available materials. For instance, floating booms can be made out of wood, bamboo, oil drums, hoses and rubber tyres, and sorbent booms constructed from fishing nets or wire mesh filled with straw, coconut husks or other indigenous materials (Figure 5a).

Figure 5a:
Improvised boom made from bamboo, rope, wire, timber and filled with rice straw sorbent.

Construc-
tion of
barriers

In shallow waters, stakes may be driven into the bottom to support screens or mats made from sacking, reeds, straw bales, bamboo or other such material (Figure 5b). On long sandy beaches bulldozers can be used to construct sand bars out into shallow water to intercept oil moving along the shoreline. A similar approach can sometimes be used to block off narrow estuaries or lagoons to prevent the ingress of oil although the ecological consequences of such temporary measures should be considered carefully. If necessary, a measure of water exchange can be achieved through pipes buried in the sand bar below low water level.

Figure 5b:
Fixed oil barrier constructed with straw bales and wire netting nailed to wooden stakes.

Bubble Barriers

A rising curtain of bubbles can be produced when air is pumped into a perforated pipe located below the water surface (Figure 6). The air bubbles create a counter-current on the water surface that holds the oil against a water flow of up to 0.7 knots. Bubble barriers are sometimes permanently installed to protect harbours where currents are relatively low and where floating booms would hinder the movement of ships. Checks of such systems are essential to ensure the air holes are not blocked by silt or marine organisms. Portable systems have also been developed which may be useful for diverting oil in calm conditions, such as in small tidal inlets or for placing across the entrance to yacht basins and marinas.

Uses

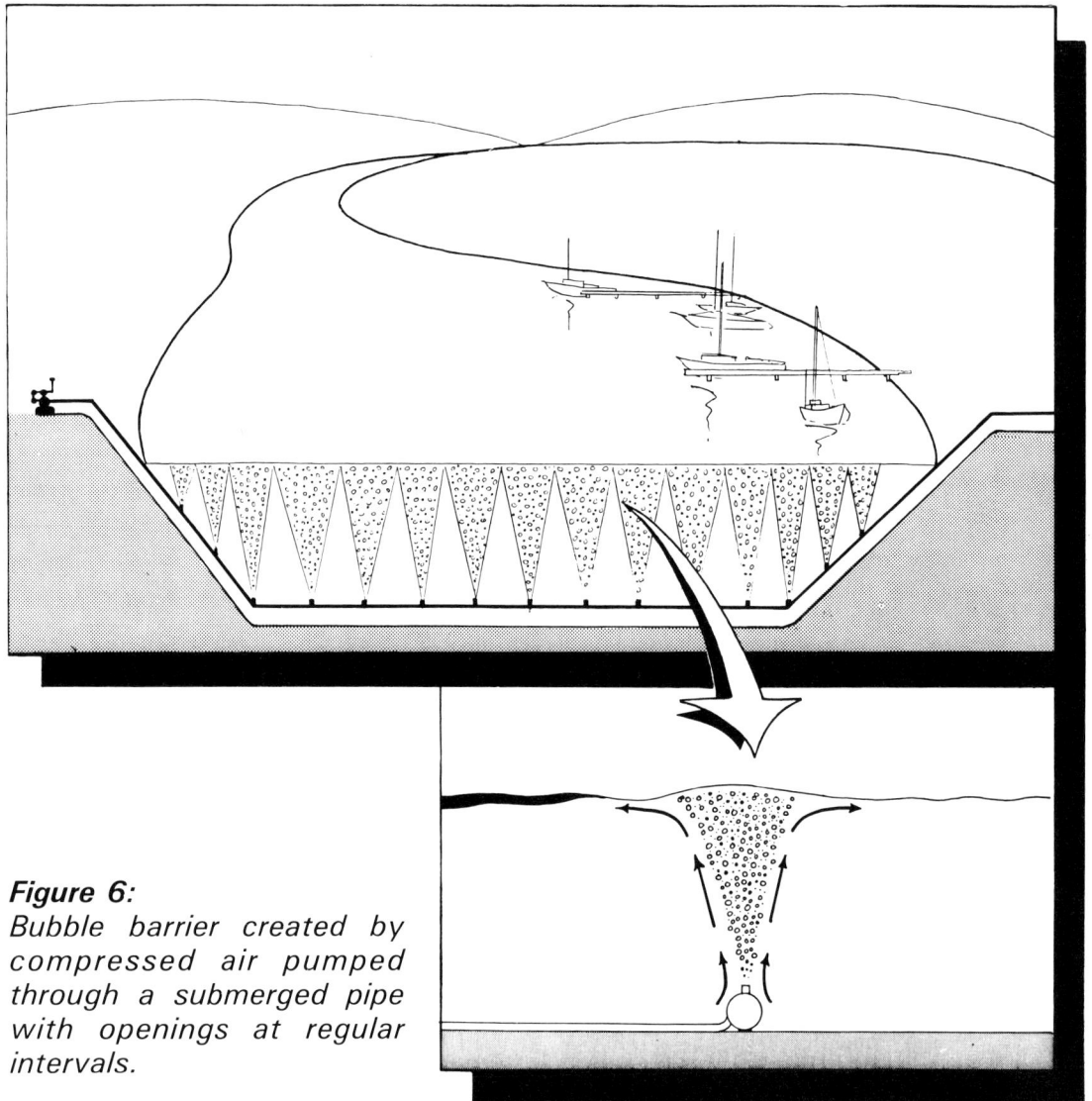

Figure 6:
Bubble barrier created by compressed air pumped through a submerged pipe with openings at regular intervals.

Chemical Barriers

Application

Certain chemicals, often referred to as "herders", inhibit the spread of low viscosity crude oils by reducing the surface tension of the surrounding water. They are designed to be sprayed at very low application rates (typically 30 litres/km around the slick perimeter) from a boat or helicopter and to surround the slick so that it will be concentrated for recovery. While the technique has been demonstrated experimentally, the high cost of these chemicals, the limited thickness of the resulting oil film and the short duration of the effect have meant that they have never been used on a large scale during an actual spill.

Limitations

RECOVERY

The rapid recovery of contained oil is vital to prevent its escape and the contamination of other areas. Recovery can be achieved using skimmers, pumps, sorbents, manual techniques and non-specialised mechanical equipment, such as vacuum trucks.

Skimmers

Design features

All skimmers incorporate an oil recovery element, some form of flotation or support arrangement and a pump to transfer collected material to storage. More complicated designs may be self-propelled and may have several recovery elements, integral storage tanks or oil/water separation facilities. A summary of the characteristic features of the main types of skimmer is given in Figure 7, overleaf.

Suction skimmers

Two basic approaches can be recognised: SUCTION and ADHESION. The simplest concept is a suction device whereby oil is collected by a pump or air suction system from the water surface directly or via a weir. These designs tend to collect large volumes of water together with the oil. This can be an advantage when recovering viscous oils since the presence of excess water helps to maintain the flow of oils which would otherwise tend to block hoses and pipework. Large storage is required to receive and separate the water which frequently represents more than 90% of the collected material. For oil spill control purposes, simple gravity separation in settling tanks is adequate.

Adhesion skimmers

Oil types

In contrast, skimmers which incorporate oleophilic materials into belts, drums, discs or synthetic ropes often achieve a higher ratio of recovered oil in relation to water. In general, they work best with medium viscosity oils between 100 and 2000 centistokes although skimmers with toothed discs or chain link belts have been designed specifically for the recovery of heavy oils. These high viscosity oils, such as heavy bunker oil, are extremely sticky and can prove difficult to remove from the adhesion surfaces, whereas, in contrast, viscous water-in-oil emulsions can be almost non-adhesive. Although low viscosity oils like diesel and kerosene can be collected, they do not accumulate on the oleophilic surfaces of skimmers in sufficiently thick layers for high recovery rates to be obtained.

Waves/swell

Currents

Skimmers are designed so that the oil recovery element is positioned at the oil/water interface. This is usually achieved by a self-levelling arrangement and although swell alone does not generally affect performance, none is effective in steep waves. Small units are easily swamped and pitched around, whilst larger skimmers have greater inertia and cannot follow the wave profiles. The performance of skimmers is also adversely affected by currents in much the same way as for booms. This limitation is partly overcome in some self-propelled skimmers where a sorbent mop array or belt is rotated so that its velocity relative to the floating oil is effectively reduced when the vessel is underway.

Self-propelled skimmers

Other designs of self-propelled skimmers can be effective in the calmer waters of ports and harbours. Because they are comparatively expensive they often combine some secondary function such as debris or waste oil collection. Such vessels are often an integral part of response arrangements for oil terminals and refineries where the pollution risk is more predictable.

Power source

Skimmers require power for the recovery element or for transferring the collected oil to a storage tank. Many systems are designed with an integral power pack. Diesel power can be used directly or to drive electric, hydraulic or pneumatic systems. All except petrol engines can be built to conform with safety regulations imposed in refineries, tank farms and other restricted areas where there may be a risk of fire and explosion. When used in potentially dangerous atmospheres, regular tests should be carried out with explosion meters to ensure safe operating conditions, since spark sources can never be completely eliminated.

Figure 7: Skimmer types

ADHESION DEVICES

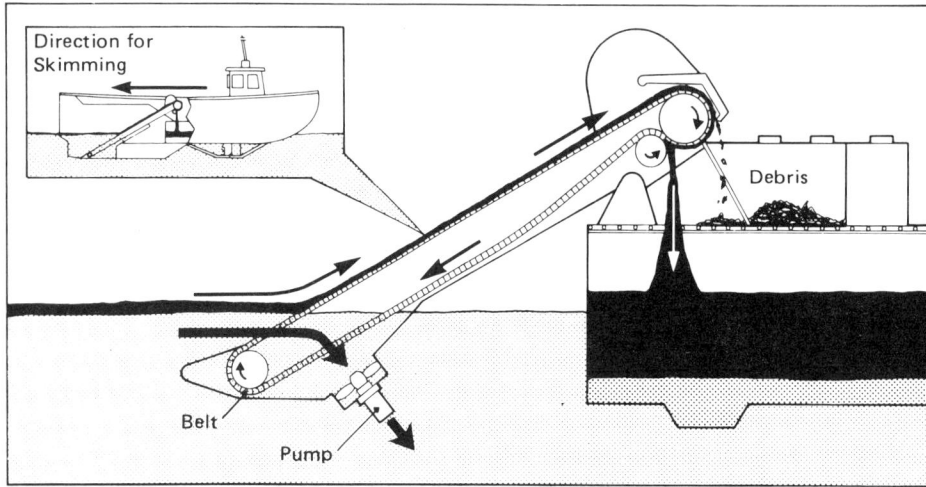

Belt skimmers

A belt conveys the oil from the water surface by adhesion. Upward rotating belts carry the oil to their top limit where it is scraped or squeezed off into a storage tank. Conversely, downward rotating belts first submerge the oil which then surfaces behind the belt, due to its buoyancy, into a defined area within the vessel. Operational limit – for upward rotating belts 0.5 knots, sea state 1; for downward rotating belts 2 knots, sea state 2. Preference – medium viscosity oils but upward rotating belts also tolerate heavier material.

Oleophilic rope skimmers

A central tension core rope, through which is interwoven oleophilic strands forming a long continuous mop. The floating mop is pulled by powered rollers around a return pulley. The rollers squeeze the oil into a storage tank. Operational limit – sea state 3. Sensitive to increasing viscosity. Preference medium viscosity oils.

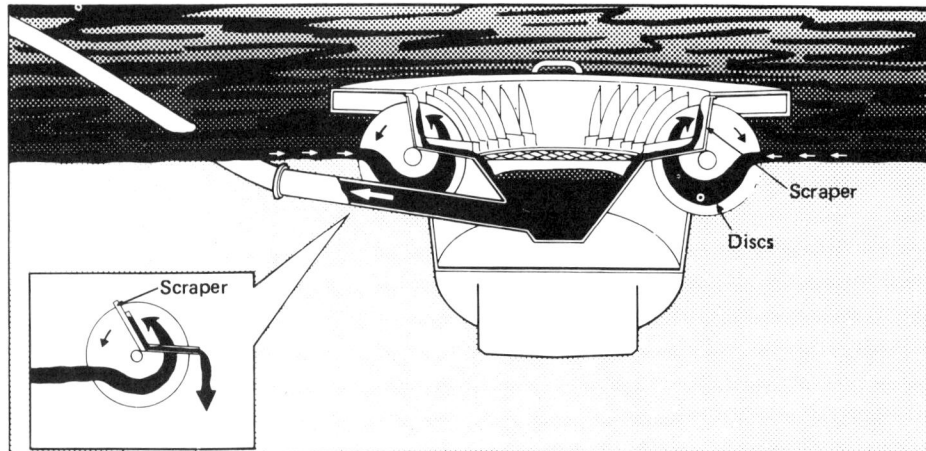

Disc skimmers

Discs rotate through the oil/water interface. Oil adheres to the disc surface, is removed by scraper to a central collection point and is pumped to storage. Operational limit – sea state 2. Sensitive to emulsified oils, waves, debris. Preference – medium viscosity oils.

SUCTION DEVICES

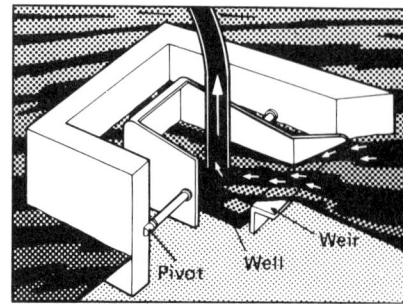

Weir skimmers

Oil flows over a self-levelling weir into the well of the skimmer and is pumped to storage. Operational limit – sea state 1. Sensitive to higher viscosity oils, emulsified oils, waves and debris. Preference – free-flowing oils.

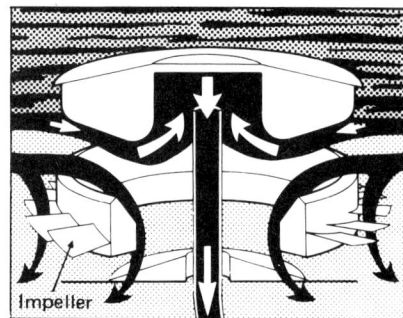

Vortex skimmers

A vortex induced by an impeller causes the oil to concentrate at the centre of the vortex due to centrifugal effects. The collected oil is pumped from the top and the free water released from the bottom. Operational limit – sea state 2 and 0.5 kt water movement. Sensitive to debris. Preference – free-flowing oils.

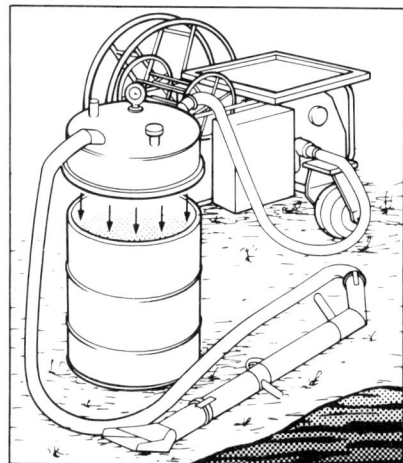

Air suction skimmers

Vacuum system or an air conveyor attached to a hose which may be fitted with specially designed skimmer heads. The pumping of more viscous materials is possible by increasing the water content. Operational limit – sea state 3. Vacuum systems more sensitive to debris. Preference – light to medium viscosity oils but air conveyors can tolerate high viscosity oils.

Unspecial-
ised
equipment

Experience has shown that the performance of skimmers is poor unless they can be deployed in calm waters and in relatively thick layers of oil. However, in these situations a simple suction device such as a vacuum truck may achieve comparable or even better recovery rates provided the suction head can be positioned in the oil layer.

Selection of
skimmers

Many factors should be considered when selecting skimmers. The intended use and expected operational conditions should first be identified before criteria such as size, robustness and ease of operation, handling and mainentance are taken into account. The most important considerations are the viscosity and adhesive properties of spilt oil, including any change in these properties over time. In predictable situations such as at marine terminals and refineries the type of oil handled is known and the optimum type of skimmer can be identified in advance.

Performance/Limitations

Pumps

The pumping phase of the skimming process often determines the overall performance of a device because all pumps lose efficiency, albeit at different rates, as oil viscosity increases. Some specialised screw pumps have a very high viscosity tolerance and can even deal with solidified oil, but the internal resistance of hoses and pipes may then become limiting instead. These problems can be overcome by entraining small amounts of oil in larger volumes of induced air or water which act as a carrier. Where viscosity increase is due to the formation of water-in-oil emulsion, chemical emulsion breakers can be employed providing they are well mixed in. Steam heating can also be useful to reduce blocking of pumps and hoses in cold conditions.

Emulsion
breakers

Hoses

Hoses carrying oil from the skimmer should be fitted with the correct flotation to prevent them interfering with the buoyancy and performance of the skimmer. All hoses can prove troublesome to handle when oily and should be fitted with effective but simple couplings. A selection of adapters can prove useful for matching hoses of different diameters and joining incompatible connectors.

Debris

Debris can also be a problem with some designs of skimmers, especially those with moving parts in direct contact with the oil. Debris can also block pumps and hoses. Some designs are fitted with coarse screens to overcome this, but the screens themselves must be kept clear if the flow of oil is not to be interrupted.

Storage of
recovered
oil

The performance of skimmers can, on occasions, be limited due to the lack of adequate or suitable temporary storage facilities. Devices that collect large volumes of water in addition to oil will, in particular, require ample temporary storage facilities to enable free water to separate and be drained off before the oil is transported to final disposal.

Recovery
rates

Because of the various constraints imposed on skimmers in the field, their design capacities are rarely achieved. Experience has consistently shown that oil cannot be concentrated sufficiently to sustain the optimum recovery rates achieved under test conditions. Test results may therefore be misleading and should be used for comparative purposes only.

Sorbents

Types

Sorbents can be defined as any material which will recover oil through absorption or adsorption. There are three basic kinds of sorbents:

1) natural organic materials such as bark, peat moss, straw, hay, feathers, coconut husks, sugar cane waste (bagasse);
2) mineral-based materials such as vermiculite, perlite and volcanic ash;
3) synthetic organic sorbents such as polyurethane foam and polypropylene fibres.

Capacity for oil	Synthetic organic sorbents usually have the greatest capacity for oil retention relative to their own volume and can be obtained in a variety of forms, including fibres, mops, sheets and pillows. They are' to some extent reusable. Some sorbents, especially natural ones, can be treated with oleophilic agents or by controlled heating which improves the ability of the material to be preferentially wetted by oil instead of water. This improves their performance and prevents them from becoming water-logged and sinking.
Uses *Recovery*	In general, the use of sorbents is only appropriate during the final stages of clean-up or to help the removal of thin films of oil from inaccessible locations. Application is normally by hand or, in the case of large scale use of loose material, by a blower. In sensitive areas such as marshes, natural sorbents may sometimes be useful to immobolise the oil and so reduce the amount available to contaminate vegetation and birds. In such situations it may be appropriate to leave the oil soaked sorbent to degrade naturally unless it is likely to migrate and contaminate other areas. However, it is usually necessary to recover all the oil soaked material so that the problems caused by the oil alone are not exacerbated. Recovery by manual means is frequently the only way since many skimmer types become clogged by sorbent materials. Exceptions are some belt skimmers and nets.

Solidifying Agents

Uses *Experience*	Although not an oil recovery technique in its own right, chemicals have been developed that convert liquid oil into solid mats, thereby facilitating recovery by manual means or nets. Whilst such chemicals have been demonstrated successfully in the laboratory, difficulties of achieving the required intensive mixing into the oil and their high cost are likely to preclude their use except for small pockets of oil in restricted locations.

Manual Recovery

Viscous oils/debris *Unspecialised equipment*	Where access is difficult, oil may have to be removed using buckets, shovels and other simple equipment. Manual recovery of large quantities of highly viscous oils and oil mixed with debris, which are not amenable to recovery by skimmers, can be aided by unspecialised mechanical equipment. For example, viscous oils and debris can be lifted out of the water using a clam shell or the bucket of a back hoe, free water spilling from between the joints in the bucket.

OFFSHORE OPERATIONS

Only rarely is it possible to use a skimmer effectively without the aid of a boom to restrict the spread of the oil over the surface of the sea and to concentrate and retain it for recovery. Deployment strategies for booms will therefore largely determine operating practices for skimmers.

Multi-vessel Operations

Boom configur- ations

Waves

In an effort to prevent spreading and maximise encounter rate, long booms of 300 metres or more may be towed in a U, V or J configuration using two vessels. The collection device is either towed together with the boom array (Figures 8a and 8b) or is deployed separately from a third vessel behind the boom (Figure 8c). The skimmer should be kept in the thickest part of the oil, yet should not be allowed to abrade or damage the boom. Wave reflection from large skimmers can interfere with the oil flow to the recovery element and skilful handling of the equipment is called for with continuous adjustments as conditions change.

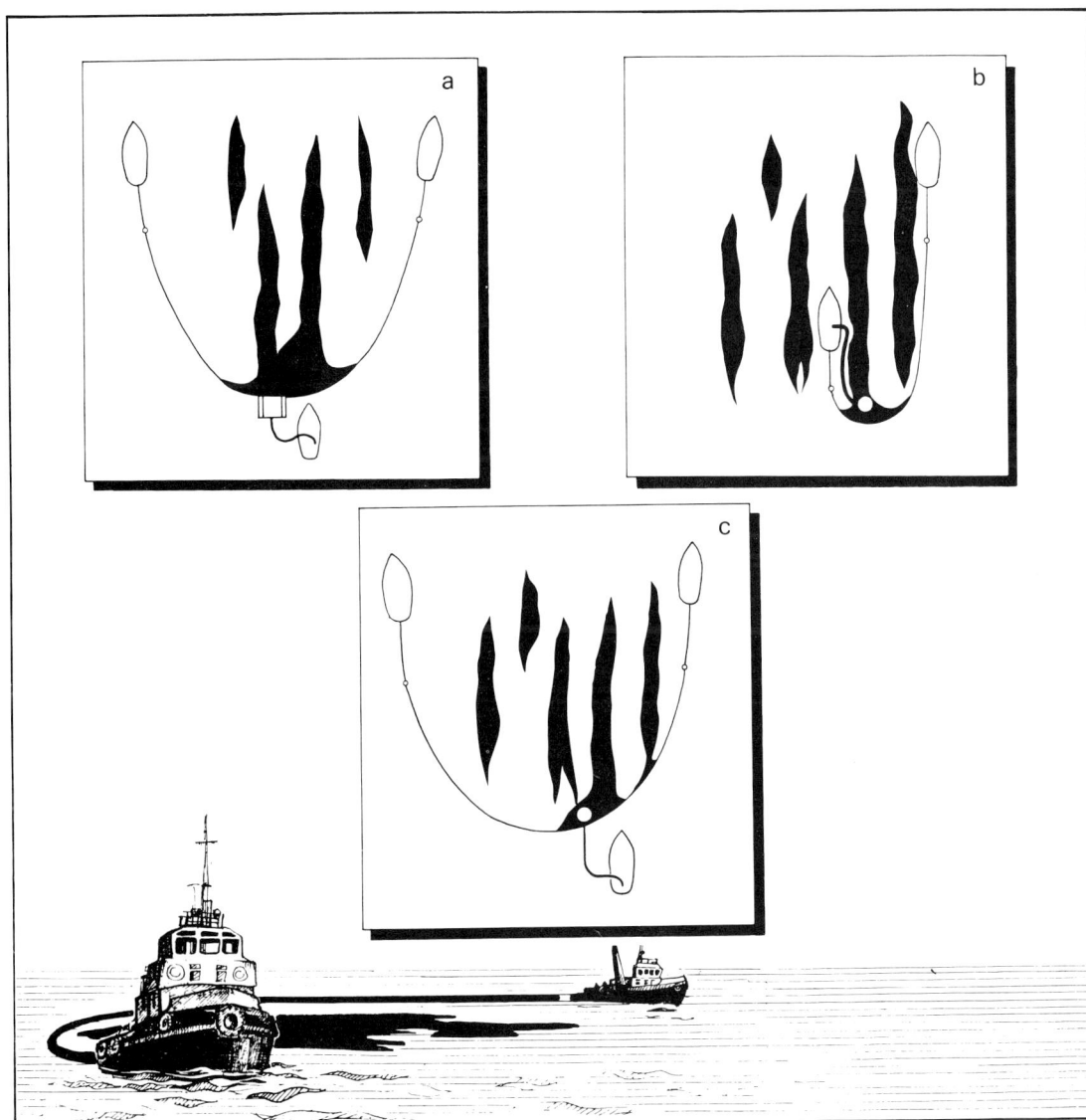

Figure 8:
a) *V configuration towed by two vessels — collection device towed with boom array and oil transferred to third vessel.*
b) *J configuration towed by two vessels, one of which deploys collection device.*
c) *U configuration towed by two vessels — collection device deployed from third vessel.*

Optimum boom design

One effective design of boom for this purpose incorporates tension members separated from the boom fabric by means of either an array of bridles or a section of netting beneath the skirt (Figure 9). These arrangements encourage the boom to remain upright under tow and leave it free to adapt to wave motion. In contrast, booms tensioned at their midpoint or waterline have a tendency to skim along the water surface.

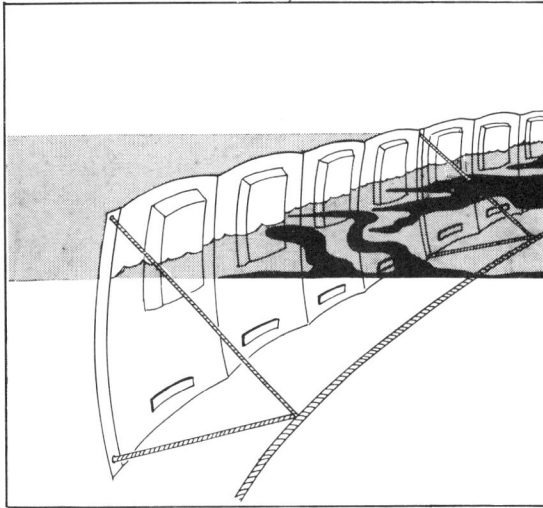

Figure 9:
Fence boom with separate tension rope attached by bridles to stiffeners in the boom face.

Towing of booms

When towing a sectioned boom in a 'U' configuration an odd number of sections of boom should be used to prevent having a join in the centre of the boom from which oil can more easily escape. To avoid sharp stress or snatching on a towed boom, lines between boom ends and the vessel should be of sufficient length. Fifty metres or more would be appropriate for towing a 300 metre length of boom.

Indications of performance

Boom performance can be judged at the apex of the 'U' or 'J' by eye. Oil lost under the boom will appear as globules or droplets rising 2-10m behind the boom. Eddies behind the boom are also an indication that it is being towed too fast. Sheens will, however, be present even when the boom is functioning well.

Vessel requirements

Specialised vessels are required to maintain both the correct configuration of the towed booms and the required very low speeds through the water, i.e. less than the escape velocity. Not only will each of the two towing vessels need at least half the total power required to tow the boom at the maximum speed consistent with oil retention, but also maximum manoeuvrability at slow speeds. As a guide, each rated horse-power of an inboard engine provides a pull of about 20 kg force. Twin propulsion units, variable pitch propellers, bow and stern thrusters are valuable. The ideal towing point aboard the vessel will need to be found by experiment and may need to be altered according to the course and wind direction. For example, a single screw vessel towing from the stern will be very unmanoeuvrable and towing from the pivoting point of the ship is preferable.

Manoeuvrability

Good manoeuvrability is just as necessary for vessels used to deploy and keep skimmers in the selected position. Lifting gear is necessary to deploy and recover the skimmer quickly. Oil concentration by towed booms can be slow and a single skimmer may be sufficient to serve several booms.

Storage of recovered oil

Recovered oil has to be stored. If the vessel used to deploy the skimmer has insufficient capacity, an additional vessel or floating flexible storage container will be needed. Removing viscous oils and water-in-oil emulsions from temporary storage tanks can create particular difficulties and should be borne in mind when selecting the temporary storage facility.

Single-ship Operations

Principal components

Co-ordinating a multi-ship operation with towed booms is very difficult and it is only rarely that any significant success has been achieved in practice. An alternative is to combine oil concentration, recovery and storage in a single ship system (Figure 10). Such systems consist of a short section of flexible boom, rigid sweeping arm or sorbent mop array to collect and concentrate the oil; a high capacity pump or skimmer to pick it up; and tankage for temporary storage of the recovered oil and water. Single ship systems operate relatively well up to sea state 3 and some can recover viscous oils, but with substantial quantities of water. Therefore onboard tank space must be adequate to allow for oil/water separation.

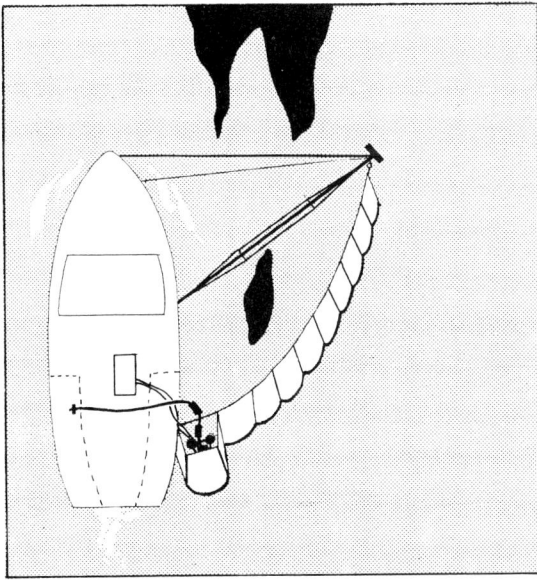

Figure 10:
Single-ship recovery system with a boom held in position by a rigid arm and guy ropes. A pocket in the boom contains a skimmer connected to a power pack on deck. Removed oil is pumped into storage tanks on board.

Perform-ance

Whilst single ship systems are more manageable than the more complex multi-ship approach, their oil encounter width is limited, being similar to the vessel's beam. This disadvantage may be less significant where floating oil is driven into narrow windrows.

Increased encounter rate

Two simple improvements can increase the oil encounter rate of single-ship systems, although both require additional vessels and equipment and so are likely to reintroduce problems of coordination (Figures 11a and 11b, overleaf). The first is for a small vessel to hold taut a relatively short length of boom attached to the collection vessel. The two vessels maintain the formation with the aim of deflecting oil into the sweeping arrangement. The second method requires two vessels towing a long length of boom at 1-2 knots. Oil that is allowed to escape behind the boom as narrow bands may then be recovered by the single ship system.

Control of Operations

Air to sea communi-cations

A common difficulty with recovery operations at sea is controlling the movements and activities of vessels and ensuring that they are operating in areas of thickest oil. This can be overcome by using aircraft equipped with suitable air to sea communications. Helicopters have the advantage of a hovering capability and can operate from a base close to the spill.

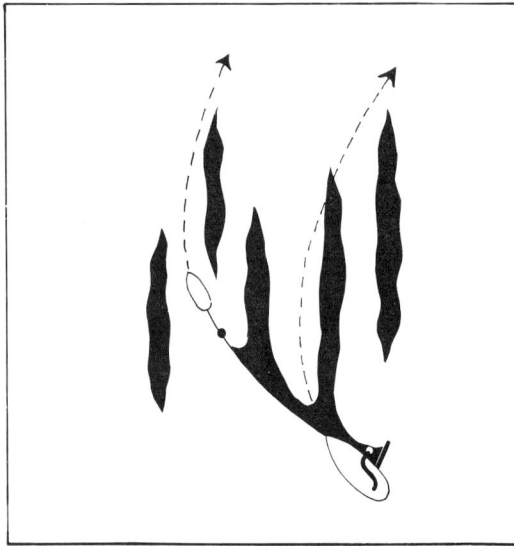

Figure 11a:
Single-ship system extended with additional vessel towing boom to increase the encounter rate.

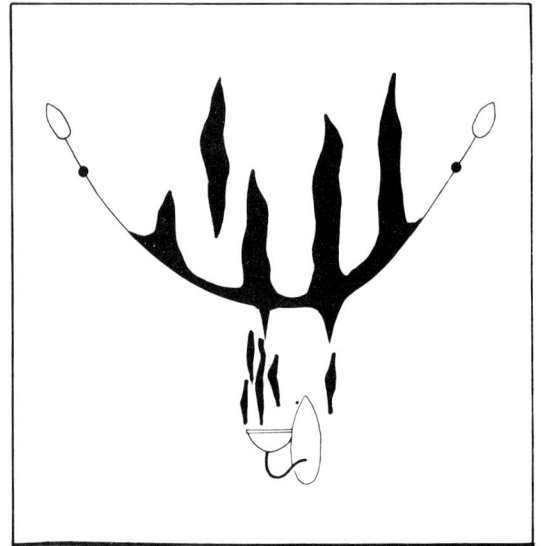

Figure 11b:
U configuration towed by two vessels at 1-2 knots. Oil escaping behind boom is intercepted by a single-ship system.

INSHORE OPERATIONS

Containment at Source

Leaking tanker/oil rig

Routine booming

Potential problems

Exceptionally it can prove practical to contain spilt oil in booms moored quite close to a source such as a leaking tanker or offshore oil well. In some ports booms are routinely placed around tankers during loading or discharging of cargo. However, it is usually difficult to keep moored booms in the desired configuration. Often waters are too exposed or currents too strong for such deployments to be effective. Also, the placing of booms close to the source may create a fire hazard and interfere with attempts to stem the flow of oil. In situations where the oil will dissipate naturally the use of booms is, in any case, inappropriate and may encourage the formation of water-in-oil emulsions.

Protection of Sensitive Areas

Priorities

Planning

More frequently, booms are used close to shore for protecting sensitive areas like estuaries, marshes, amenity areas and water intakes. In practice it may not be possible to protect all such sites. Careful planning should therefore be devoted to identifying first, those which can be boomed effectively, and second, an order of priority. A further advantage of such planning is that preparations can be made for boom selection, mooring and oil collection. Planning is particularly relevant for oil terminals and similar installations where the source of the spill and most likely size and type of oil can be predicted.

Selection of suitable sites

Removal of contained oil

An aerial survey can be valuable in identifying suitable sites for using booms. In selecting the location and method of deployment it may be necessary to compromise between conflicting requirements. For instance, although it may be desirable to protect a complete river, the estuary may be too wide or the currents too strong to achieve this. A more suitable location may have to be sought further upstream, bearing in mind the need for access to deploy the boom and remove the collected oil. If the oil is not removed at the rate of its arrival, it will accumulate and move out towards the centre of the river where the stronger currents may sweep the oil under the boom. If current velocities are unknown they can be estimated by timing the movement of floating objects over a known distance.

It is frequently better to use booms to deflect oil to relatively quiet waters where it may be recovered rather than attempt containment. As shown in Table 1 it is feasible to deflect floating oil even in a 3 knot current (1.5 m/s) where a boom positioned at right angles to the flow would fail to contain any oil. Following this principle a river can be protected by placing a boom obliquely to the direction of flow. If it is necessary to maintain a navigation channel, two sections of boom can be staggered from opposite banks (Figure 12).

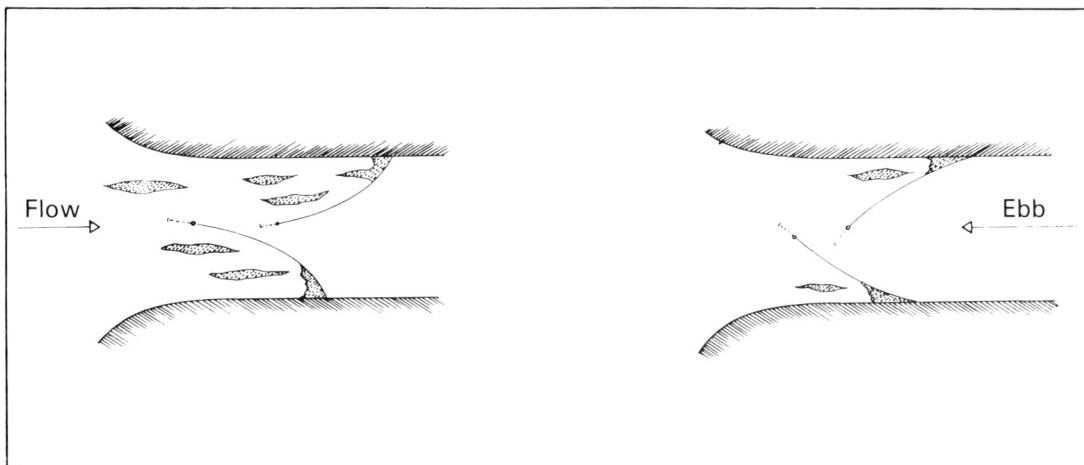

Figure 12:
Staggered boom configuration across a channel.

Table 1. Maximum boom deployment angles to flow direction at different current strengths to prevent escape of oil. Calculations based on an escape velocity of 0.7 knots (0.35 m/s).

Current strength		Max. angle
(knots)	(m/s)	(degrees)
0.7	0.35	90
1.0	0.5	45
1.5	0.75	28
2.0	1.0	20
2.5	1.25	16
3.0	1.5	13

Table 2. Holding strength of Danforth type anchors in loose mud, sand or gravel, and clay.

Anchor weight (kg)	Holding strength (kg force)		
	Mud	Sand	Clay
15	200	250	300
25	350	400	500
35	600	700	700

Diversion of Oil

Another way to protect sensitive areas where strong currents are likely to defeat containment and collection of oil, is to use booms to divert the oil. However, this is almost invariably a temporary solution and attention will need to be given to where the oil will then go, to ensure that greater problems will not be caused.

Mooring of Booms

Currents

Anchoring

Concrete blocks

Permanent moorings

Correct mooring is crucial since performance is dependant upon the angle of deflection being appropriate for the prevailing current strength. To maintain this angle and prevent the formation of pockets, several anchoring points may be required. However, to lay multiple moorings is a skilled job and often impractical in an emergency. The formula given earlier can be used together with Tables 1 and 2 as a guide to the moorings required to hold a boom in a current of known strength and likely maximum wind. A Danforth type anchor is effective on sand and mud substrates but a Fisherman's type anchor is better on rocky bottoms (Figures 13a and 13b). If there is time, concrete blocks can be cast to give convenient and reliable mooring points, but their weight in air must be at least three times the expected load, to compensate for their reduced weight in sea water. A workboat with lifting gear would be required to handle heavy moorings. In high risk areas such as near oil terminals, permanent mooring points may be provided at key places. Appropriate lengths of boom may also be stored under cover at these sites, on reels or in other ways that will facilitate rapid deployment.

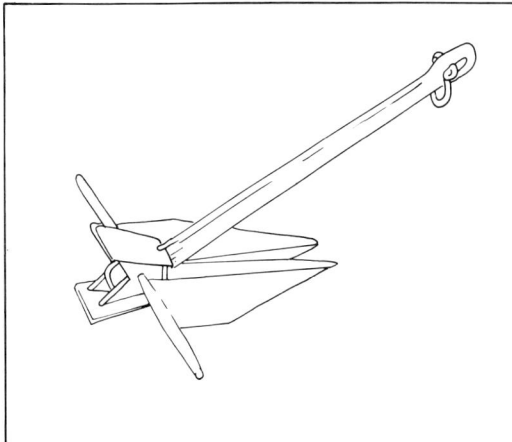

Figure 13a:
Danforth anchor.

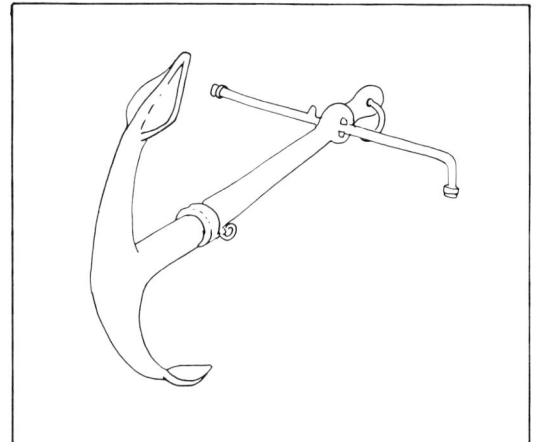

Figure 13b:
Fisherman's anchor.

Water depth, swell, tides

Mooring practices

Shorelines

Whichever type of mooring is used, it is important to select the length of the mooring lines to suit the expected water depth, swell and tidal range. If the lines are too short the boom will not ride well in the water and the snatching produced in the lines by waves may dislodge the moorings. Conversely, if the lines are too long it will be difficult to control the configuration. A length of heavy chain or other weight between anchor and line greatly improves the holding power of an anchor, and the use of an intermediate buoy between the boom and anchor will help prevent submersion of the boom at anchoring points and can absorb some 'snatch' loads. Equally, a weight hung from a mooring line stops it floating on the surface when slack (Figure 14). When deploying a boom from a shoreline it is often possible to secure it to fixed objects on the shore. On a featureless sandy beach a buried object such as a log provides an excellent mooring point.

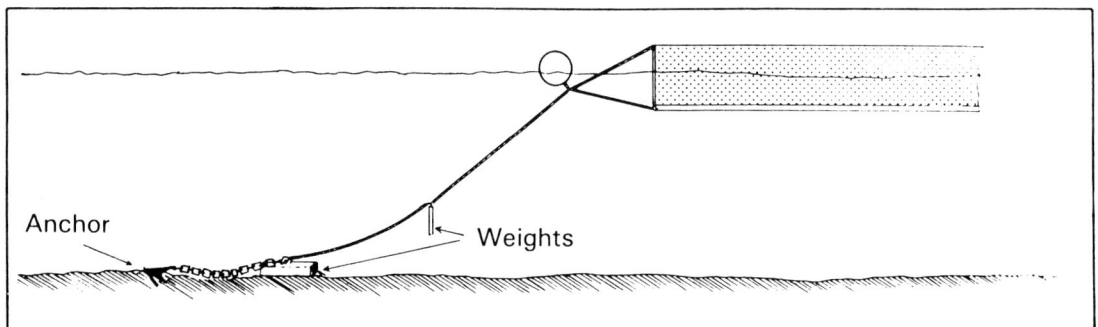

Figure 14:
Typical mooring arrangement.

In tidal areas where booms are permanently installed or where frequent deployment is likely to be required, some form of sliding mooring may be needed to provide a seal between the boom and shore at all states of the tide (Figure 15).

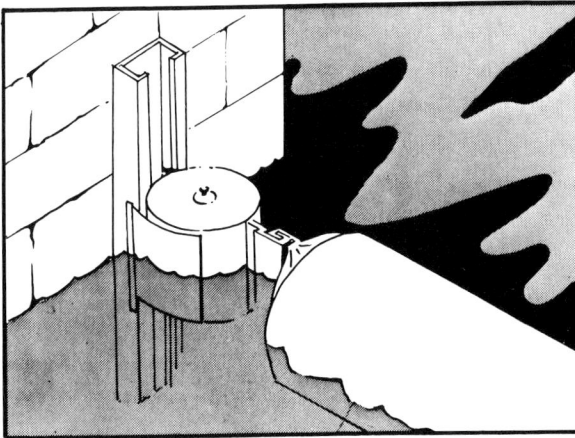

Figure 15:
Boom connection consisting of a flotation cylinder with a sliding shoe fitting into a vertical runner fixed to a harbour wall.

The use of booms on mud and sand flats in estuaries or in other situations where the boom may become stranded at low tide can cause particular problems and requires the use of types designed specifically for such situations (Figure 16).

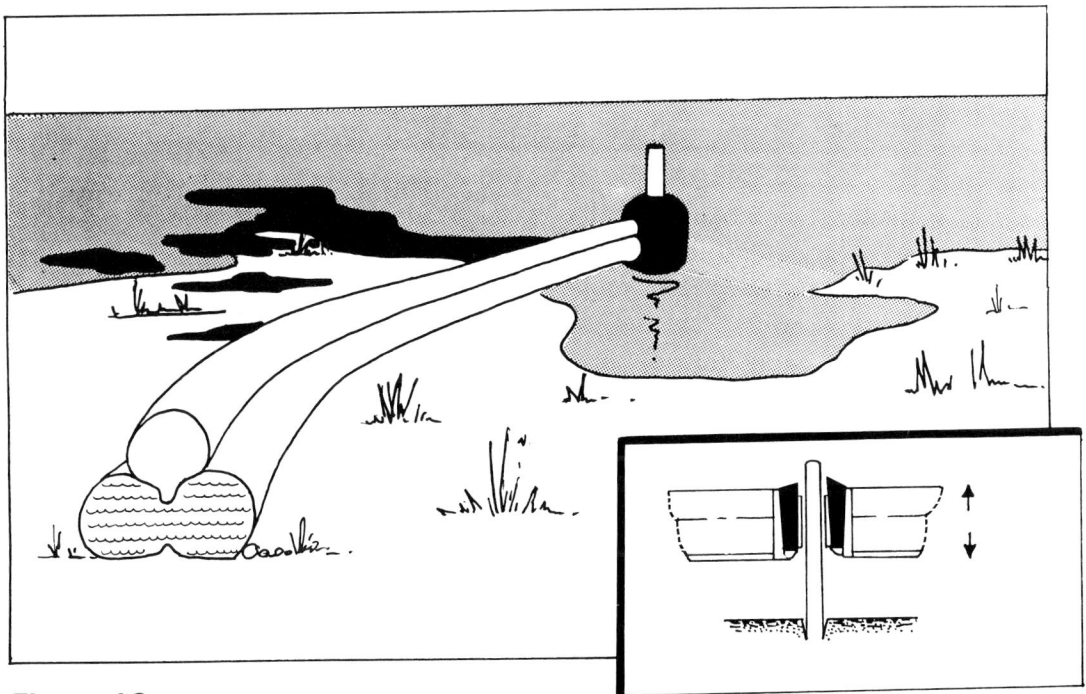

Figure 16:
Shoreline barrier with twin water tubes providing ballast underneath an air flotation tube. A hollow buoyancy chamber fitted over a pile driven into the mud gives an anchoring point which compensates for tidal movement.

Other Uses of Booms

In common with other floating materials, oil accumulates in particular places along the shore under the influence of wind and water movement. Booms can be used to further concentrate the oil and prevent it moving elsewhere if conditions change. This not only minimises the extent of the contamination but permits the controlled removal of the trapped oil. As debris is likely to be associated with the oil in such locations, skimmers must be used that are capable of operating in the presence of such material. It may be

Recovery of oil	necessary to work from shallow-draught vessels but it is usually easier to operate skimmers from the shore particularly if road access, a hard standing or a flat working area is available close to the oil collection point. Once again, the problems of temporary storage of recovered oil should not be overlooked.
Beach washing	Booms can also assist shoreline clean-up by containing oil washed off beaches and rocks. By drawing in the boom the oil can be concentrated and moved towards collection devices. In some circumstances, simple expendable sorbent booms can be used to collect thin oil films.

Inspection and Maintenance

Winds, currents, tides ***Temperature*** ***Damage by vessels*** ***Debris*** ***Optimum performance***	As winds, currents and tides change, so will the configuration of a boom. Frequent checks and re-adjustment of the moorings will be necessary. Contained oil and debris must be removed promptly since the performance of the boom will otherwise be severely reduced. In conditions where the air temperature is hot by day and cool by night it is important to allow for the expansion and contraction of air in inflatable booms. This may necessitate letting air out during the day and re-inflating at night, and it is advisable to take some precautions such as notifying mariners and displaying warning lights. Brightly coloured booms are more visible in daylight and are better picked out by lights at night.

Skimmers also require supervision to ensure that oil is reaching the collection element and that debris is not accumulating to reduce efficiency or cause damage. Many skimmers are fitted with trash screens which can become blocked by oil or debris with surprising frequency. Repairs will usually require trained personnel and a stock of replacement parts. At night, regular inspections must be made and when necessary the skimming equipment operated using floodlighting. To maintain a high performance the skimming rate should be adjusted to suit the conditions, the type of oil and to match the rate at which oil is arriving at the collection site. If only small amounts of oil are present, skimming should be done at intervals to avoid excessive collection of water. Rope mops can be used effectively inside a boom to collect small quantities of oil along its length.

Cleaning and overhaul	After use skimmers should be cleaned and overhauled to identify and rectify any wear and damage. Steam lances or hydrocarbon solvents can be used to remove oil but dispersants or detergents should not be used on sorbent mops and plastic discs which can be cleaned with diesel. Skimmers and their power packs should be protected from damage and damp salty atmospheres causing corrosion. Sorbent mops, rubber belts and plastic materials incorporated in skimmers will perish if exposed to direct sunlight for long periods. Regular inspections and testing of equipment are essential particularly as its use may be infrequent and there is a tendency for vital parts to go missing.
Storage ***Repairs***	Proper retrieval, maintenance and storage of booms away from direct sunlight is also vital to prolong their life and ensure that they are always ready for use at short notice. Booms usually require cleaning after use and this can prove difficult with some designs. Dispersants or steam cleaning are usually employed but when using chemicals it is important to ensure first that the boom fabric will not be damaged. Emergency repair kits should be kept on hand for dealing with minor damage which could otherwise make a section or even the whole length of boom unusable.

CONTINGENCY PLANNING AND TRAINING

Rapid response

When a decision is made to contain or to attempt to protect sensitive areas, a rapid response is essential. Equally important is the prompt removal of any contained oil. As noted elsewhere in this section, this can be aided by careful planning before the event, such as by: selection of priority sites for booms, preparation of suitable anchoring points, selection of skimmers to match likely requirements, provision of temporary storage facilities for recovered oil and other logistic support arrangements.

Exercises

Training of personnel, and practical exercises, are essential. Even the mobilisation and deployment of equipment under controlled conditions and in fair weather without oil will test planned procedures and strategies and will highlight some of the difficulties that are likely to be encountered in a real spill. Practical exercises will also provide local clean-up crews with an opportunity to learn new skills and to rehearse basic principles of seamanship.

POINTS TO REMEMBER

1. Assess the merits of oil containment and recovery in relation to other clean-up options and against prevailing conditions such as sea state, wind, water movements and the location of areas needing protection.

2. Decide whether selected priority coastal areas are to be protected by towed or moored booms or some other form of barrier.

3. Select suitable booms or other barriers to match requirements and prevailing conditions, taking into account availability, reliability, ease and speed of deployment and cost.

4. Select appropriate skimmers on the basis of the type and amount of oil to be recovered, its viscosity at ambient temperatures and any change with time, and taking into account availability, reliability, robustness, field performance, weight, ease of handling, power source and cost.

5. Identify the availability of vacuum trucks and other suction systems which may be employed for recovering thick layers of oil on calm waters, and other recovery techniques which may be appropriate in certain situations.

6. Use aircraft or helicopters to monitor and control operations at sea.

7. Monitor boom function and skimmer performance continuously to ensure optimum efficiency.

8. Ensure that the logistics of mobilisation, transport, deployment and storage are addressed as well as arrangements for the disposal of recovered oil.

9. Thoroughly train personnel and maintain their standards by practical exercises. Arrange suitable storage, and regular inspections and testing of equipment and rectify any faults.

10. Appreciate the limitations of equipment and materials available for containment and recovery operations and be prepared to improvise when the need arises.

FURTHER READING

CONCAWE (1981) A field guide to coastal oil spill control and clean-up techniques. CONCAWE, The Hague, Netherlands, Report No. 9/81. 112 pp.

CONCAWE (1983) A field guide to inland oil spill clean-up techniques. CONCAWE, The Hague, Netherlands, Report No. 10/83. 104 pp.

Cormack, D. (1983) Response to oil and chemical marine pollution. Applied Science Publishers, London. 531 pp.

Dhennin, A. (1983) Les barrages: La lutte contre les pollutions accidentelles des eaux. CEDRE, Brest, France. Publication No. R83.810.E. 16pp.

Environment Canada (1984) A winter evaluation of oil skimmers and booms. Environmental Protection Service, Environment Canada. Report No. EPS 4-EP-84-1. 109 pp.

Fingas, M.F., Duval, W.S. and Stevenson, G.B. (1979). The basics of oil spill clean-up. Environmental Protection Service, Environment Canada. 155 pp.

IMO (1980) Manual on oil pollution — Section IV. Practical information on means of dealing with oil spillages. IMO, London. 143 pp. (Under revision).

ITOPF (1981) Use of booms in combating oil pollution. Technical Information Paper No. 2, ITOPF, London. 8 pp.

ITOPF (1983) Use of skimmers in combating oil pollution. Technical Information Paper No. 5, ITOPF, London. 7 pp.

Koops, W. (1985) Manual of oil pollution, on sea, on coast and on inland waters. Staatsuiedecery (Government Publishing Office), The Hague, Netherlands (In Dutch).

Peigne, G. (1984) Skimmers. CEDRE, Brest, France. Publication No. R84.903.E. 60pp.

Schulze, R. (1987) World catalog of oil spill response products — Complete listing of oil spill booms, skimmers and sorbents. Robert Schulze (Ed.). Maryland, USA. 470 pp.

Solsberg, L.B. (1983) A catalogue of oil skimmers. Environmental Protection Service, Environment Canada. Report No. EPS 3-EP-83-1. 258 pp.

III   THE USE OF DISPERSANTS

The action of waves on oil slicks can promote the natural dispersion of oil into small droplets. As the droplets become mixed through the water column, the oil concentration reduces and the oil becomes more readily available for eventual degradation by micro-organisms. In order to accelerate this process it is sometimes appropriate to use a chemical dispersant, especially when containment and recovery is impractical. The removal of oil from the sea surface prevents the formation of persistent water-in-oil emulsions and residues, both of which can present a threat to coastlines and seabirds.

This section explains how dispersants work; the types available; the methods of application at sea and on shorelines and the environmental considerations involved.

CONTENTS

CHARACTERISTICS OF DISPERSANTS

Mechanism of Chemical Dispersion

Surfactants

The key component of a dispersant is a surface-active agent (surfactant) which has a molecular structure such that one part of the molecule has an affinity for oil (oleophile) and the other has an affinity for water (hydrophile). When evenly applied and mixed into floating oil the molecules become arranged at the oil-water interface in such a way that the interfacial tension between oil and water is reduced. This promotes the formation of finely dispersed oil droplets with a combined surface area much greater than the original oil slick (Figure 1). For a fixed amount of mixing energy, a reduction in interfacial tension between oil and water will result in a corresponding increase in combined droplet area.

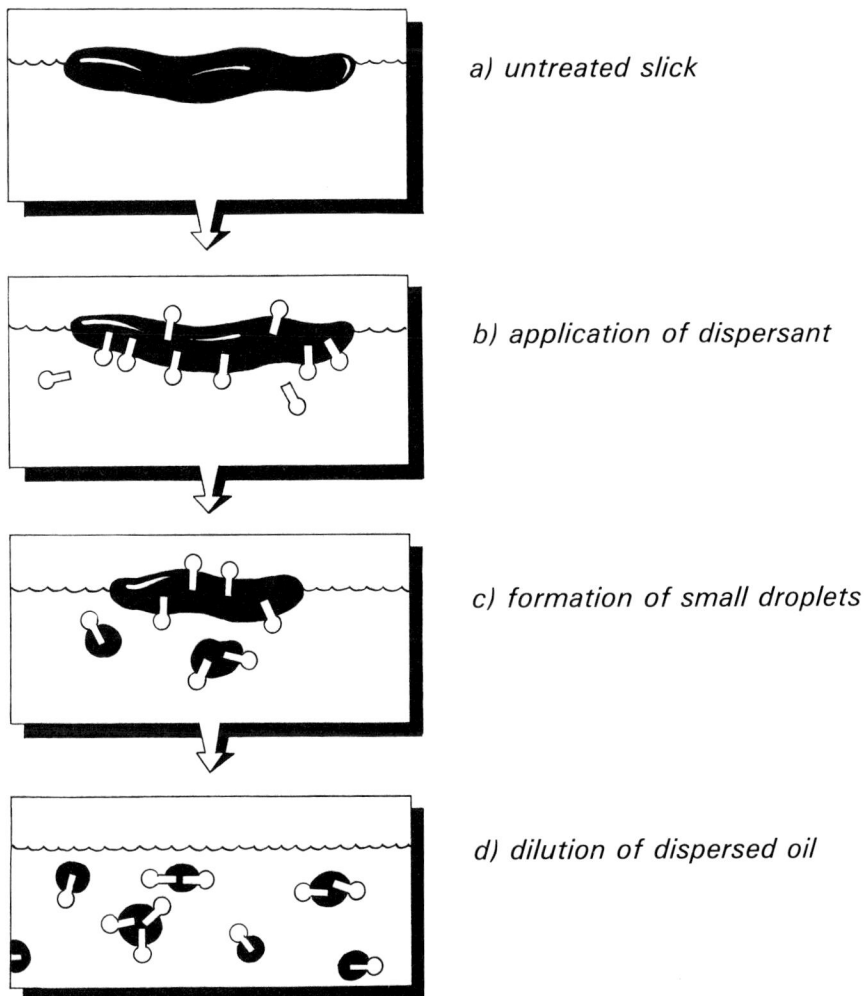

a) untreated slick

b) application of dispersant

c) formation of small droplets

d) dilution of dispersed oil

Figure 1: Mechanism of dispersion

Oil droplet size

Appearance of dispersed oil

The smaller the droplets, the greater the chance that they will remain suspended in the water because of their very slow rise velocity. In practice, for most crude oils, provided the droplets produced have an average diameter of 0.2 mm or less, they are unlikely to return to the sea surface except in calm conditions. However, in the case of light refined oil products with low specific gravities, such as gas oil or diesel, much smaller droplets are necessary to overcome their greater buoyancy. If a black oil is satisfactorily dispersed, a characteristic coffee-coloured cloud or plume can be seen to spread slowly down from the water surface a few minutes after treatment.

*Prevention
of
coalescence*

As well as promoting droplet formation, dispersants perform a secondary role by preventing coalescence of the oil droplets once they are formed. This is because the surface-active agent remains at the oil-water interface long enough to act as a barrier between any droplets which collide with one another at random.

Solvents

Extensive laboratory and field testing of the effectiveness of different dispersants on different oils has shown that for a dispersant to work it must be distributed throughout the oil. Most dispersants therefore contain a suitable solvent or combination of solvents which penetrate the oil and act as a carrier for the surfactant.

Dispersant Effectiveness

*Maximum
viscosity*

*Pour
point*

If the oil is very viscous dispersants are ineffective since they tend to run off the oil into the water before the solvent can penetrate. As a general rule, dispersants are capable of dispersing most liquid oils and liquid water-in-oil emulsions with viscosities less than about 2000 centistokes (cSt), equivalent to a medium fuel oil at 10-20°C. Once viscosities exceed 2000 cSt., dispersants applied at sea become less effective and as viscosities reach the region of 5,000-10,000 cSt., they are ineffective. They are not suitable for dealing with viscous emulsions (mousse) or oils which have a pour point near to or above that of the ambient temperature. Heavy fuel oil is seldom dispersible and Table 1 lists some crude oils which may not be amenable to dispersants due to their viscosity or pour point and Figure 2 gives some typical viscosity values for different oils at various temperatures. Even those oils which are dispersible when spilled, become resistant because of increased viscosity due to weathering.

*Rapid
response*

The time taken for evaporation and emulsification to render a particular oil resistant to dispersant depends upon such factors as sea state and temperature but in most cases it is unlikely to be longer than a day or two. This means that response teams deciding to apply dispersant must act quickly. Dispersants are sometimes more effective with viscous oils stranded on shorelines because the contact time can be prolonged to allow penetration of the chemical into the oil layer.

Types of Dispersant

Dispersants in use today are of two main types:

*Hydro-
carbon
products*

A. Hydrocarbon or conventional dispersants are based on hydrocarbon solvents and contain between 15 and 25% surfactant. They are intended for neat application to oil and should not be diluted with sea water prior to application since this renders them ineffective. Typical dose rates are between 1:1 and 1:3 (dispersant:oil). Early hydrocarbon formulations, sometimes referred to as first generation, used toxic aromatic solvents whereas present-day products, known as second generation, have an aromatic-free solvent.

*Concen-
trates*

B. Concentrate or self-mix dispersants have alcohol or glycol solvents and usually contain a higher concentration of surfactant components. These third generation products should preferably be applied neat but can be diluted with sea water before spraying. Typical dose rates are between 1:5 and 1:30 (neat dispersant:oil).

Mixing

Concentrates diluted with sea water and hydrocarbon-based dispersants both require thorough mixing with the oil after application to produce a satisfactory dispersion. However, concentrates, if sprayed undiluted directly on to the oil, do not require the same degree of mixing and usually the wave action is sufficient. As a result of this, and the lower application rates required, concentrates have largely superseded hydrocarbon-based dispersants for application at sea.

TABLE 1. Crude Oils with High Viscosities or High Pour Points

The pour point of an oil is the temperature below which the oil will not flow. The oils set in **bold** type are unlikely ever to be amenable to dispersants because of high viscosities or high pour points. The other oils listed could be amenable to dispersants when the ambient temperature is high. To determine viscosities at ambient temperatures, refer to Figure 2, note 1.

Crude Name	Loading Port	Country	Gravity* °API	Viscosity cSt 100°F 37.8°C	Pour point °C	Pour point °F
Amna	Ras Lanuf	Libya	36.1	13	18	65
Ardjuna	SBM	Indonesia, E. Kalimantan	37.7	3	27	80
Bachequero	La Salina	Venezuela	16.8	275	−23	−10
Bahia	**Salvador**	**Brazil**	**35.2**	**17**	**38**	**100**
Bakr	Ras Gharib	Egypt	20.0	152	7	45
Bass Strait		Australia	46.0	2	15	60
Belayim	Wadi Feiran	Egypt	27.5	18	6	43
Boscan	**Bajo Grande**	**Venezuela**	**10.3**	**>20,000**	**15**	**60**
Bu Attifel	**Zueitina**	**Libya**	**40.6**	**—**	**39**	**102**
Bunju	Balikpapan	Indonesia, E. Kalimantan	32.2	3	17.5	63
Cabinda	SPMB-Landana	Angola	32.9	20	27	80
Cinta	**SBM**	**Indonesia, Sumatra**	**32.0**	**—**	**43**	**110**
Duri	Dumai	Sumatra	20.6	100	14	57
El Morgan	Shaukeer	Egypt	32.3	9.5	7	45
Es Sider	Es Sider	Libya	37.0	5.7	9	48
Gamba	SPMB-Gamba	Gabon	31.8	38	23	73
Gippsland Mix	Western Port Bay	Australia	44.4	2	15	60
Handil	**SBM**	**Indonesia, E. Kalimantan**	**33.0**	**4.2**	**29**	**85**
Heavy Lake Mix	La Salina	Venezuela	17.4	600	−12	10
Iranian Nowruz	Bahrgan	Iran	18.3	270	−26	−15
Jatibarang	**SBM**	**Indonesia, Java**	**28.9**	**—**	**43**	**110**
Jobo/Morichal (Monagas)	**Puerto Ordaz**	**Venezuela**	**12.2**	**3780**	**−1**	**30**
Lagunillas	La Salina	Venezuela	17.7	500	−20	−5
Mandji	Cap Lopez	Gabon	29.0	17	9	48
Merey	Puerta La Cruz	Venezuela	17.2	520	−23	−10
Minas	**Dumai**	**Indonesia, Sumatra**	**35.2**	**—**	**32**	**90**
Panuco	**Tampico**	**Mexico**	**12.8**	**4700**	**2**	**35**
Pilon	**Carpito**	**Venezuela**	**13.8**	**1900**	**−4**	**25**
Qua Iboe	SBM	Nigeria	35.8	3.4	10	50
Quiriquire	Carpito	Venezuela	16.1	160	−29	−20
Ras Lanuf	Ras Lanuf	Libya	36.9	4	7	45
Rio Zulia	Santa Maria	Colombia	40.8	4	27	80
San Joachim	Puerto La Cruz	Venezuela	41.5	2	24	75
Santa Rosa	Puerto La Cruz	Venezuela	49.4	2	10	50
Seria	Lutong	Brunei	36.9	2	2	35
Shengli	Qingdao	P.R. China	24.2	—	21	70
Taching	**Darien**	**P.R. China**	**33.0**	**138**	**35**	**95**
Tia Juana Pesada	**Puerto Miranda**	**Venezuela**	**13.2**	**>10,000**	**−1**	**30**
Wafra Eocene	Mina Saud/Mina Abdulla	Neutral Zone/Kuwait	18.6	270	−29	−20
Zaire	SBM	Zaire	34.0	20	27	80
Zeta North	Puerto La Cruz	Venezuela	35.0	3	21	70

* $°API = \dfrac{141.5}{\text{Specific Gravity}} - 131.5$

* $\text{Specific Gravity} = \dfrac{141.5}{°API + 131.5}$

Figure 2: *Relationship between temperature and oil viscosity for representative crude and fuel oils.*

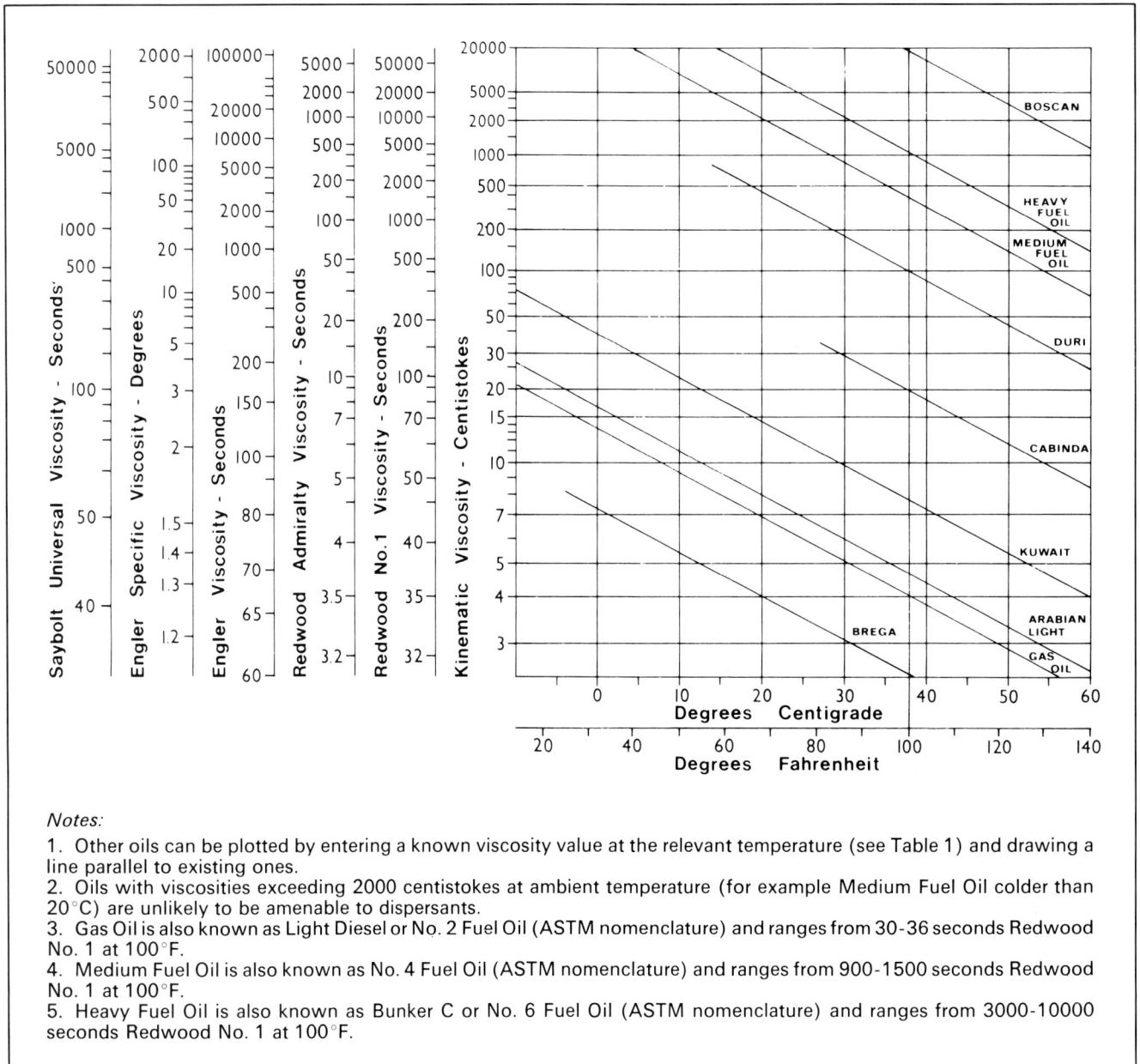

Notes:
1. Other oils can be plotted by entering a known viscosity value at the relevant temperature (see Table 1) and drawing a line parallel to existing ones.
2. Oils with viscosities exceeding 2000 centistokes at ambient temperature (for example Medium Fuel Oil colder than 20°C) are unlikely to be amenable to dispersants.
3. Gas Oil is also known as Light Diesel or No. 2 Fuel Oil (ASTM nomenclature) and ranges from 30-36 seconds Redwood No. 1 at 100°F.
4. Medium Fuel Oil is also known as No. 4 Fuel Oil (ASTM nomenclature) and ranges from 900-1500 seconds Redwood No. 1 at 100°F.
5. Heavy Fuel Oil is also known as Bunker C or No. 6 Fuel Oil (ASTM nomenclature) and ranges from 3000-10000 seconds Redwood No. 1 at 100°F.

METHODS OF APPLICATION AT SEA

The method of application depends primarily on the type of dispersant, the size and location of the spill, and the availability of vessels or aircraft for spraying the dispersant. Table 2 summarises the main characteristics of various dispersant spraying systems.

TABLE 2. Comparison of Dispersant Application Systems

Spraying system	Type of dispersant	Maximum dispersant application rate litres/min.	Maximum oil treatment rate tonnes/hr.	Advantages	Disadvantages
BACK PACK UNITS	HYDROCARBON	2.5	0.3	Light, portable, cheap and readily available	Limited payload and application rate
	CONCENTRATE	2.5	3		
FIRE MONITORS	CONCENTRATE	10-70	1	Available on most vessels	Limited oil encounter rate, and wastage of chemical dispersant
WARREN SPRING OFFSHORE SPRAYING EQUIPMENT	HYDROCARBON	90	10	Relatively cheap, can be fitted to most types of vessel	Limited oil encounter rate, cumbersome surface agitation boards, cannot be mounted on bow to avoid bow wave effect. Constant pumping rate
	CONCENTRATE	9.0	10		
WARREN SPRING INSHORE SPRAYING EQUIPMENT	HYDROCARBON	32	4	Relatively cheap, readily fitted to most vessels with 15 hp engine and can be adapted for shore clean-up	Limited oil encounter rate, cannot be bow mounted to avoid bow wave effect. Constant pumping rate
	CONCENTRATE	3.2	4		
SPRAYING EQUIPMENT FOR NEAT CONCENTRATE	CONCENTRATE	220	70*	Relatively cheap, can be fitted to bow of most vessels and provides variable application rate	Very high potential for dispersant wastage due to high application rate
PIPER PAWNEE CROP SPRAYING AIRCRAFT	CONCENTRATE	120	40	Rapid response, high treatment rate, accurate distribution of chemical particularly for fragmented slicks. Can land and take off from rudimentary airfields	Limited payload and endurance. Single engined therefore only suitable for inshore waters
DOUGLAS DC6 SPRAYING AIRCRAFT	CONCENTRATE	400	320	Rapid response, very high treatment rate, only suitable for very large spills	Requires costly dedicated aircraft, time to position at spill site, long runway and controlling aircraft

*Based on a typical oil encounter rate for a vessel travelling at 10 knots and spraying a 40 metre swath.

Vessel Spraying

Size of vessel

Vessels used in spraying operations vary considerably in size, ranging from large purpose-built ships to small vessels of opportunity onto which spraying equipment is installed. Small vessels are not very suitable for major operations since they do not have sufficient storage capacity for the large quantities of dispersant required. The use of concentrate dispersants makes the best use of the limited capacity of such vessels and permits the treatment of a greater volume of oil than the same quantity of hydrocarbon dispersant. Once the oil moves away from the source of the spill, vessels of all sizes need the assistance of an aircraft overhead to direct them into the slick.

Improvised Spraying Equipment

Fire pumps

When specialised spraying equipment is not available, fire pumps and hoses are sometimes employed as a last resort. However, hydrocarbon solvent-based dispersants should never be used with this equipment, since the pre-dilution with water inactivates the surfactant. Concentrate dispersant is drawn into the sea water stream by an eductor but the output must be controlled by the pumping rate or by bleeding off excess water to achieve a 10% concentration of dispersant. Because the water flow rates tend to be very high, both excessive dilution and high consumption of chemical may result. Efficient application *Limitation* is also difficult because only a small area is covered by the strong jet of water produced by a fire hose. Despite these limitations, fire hoses or fire monitors can be suitable for dealing with small spills in confined spaces, such as under jetties.

Specialised Spraying Equipment

Surface agitation boards

When hydrocarbon-based dispersants or diluted concentrates are applied from spray booms mounted on vessels, it is possible to achieve the required mixing by towing surface agitation boards through the treated slick (Figure 3a). With this equipment, the dispersant is usually sprayed at a constant rate and the ratio of dispersant to oil can only be adjusted by varying the speed of the vessel or closing off one of the spray booms. In practice, the vessel can only travel between 4 and 10 knots, as this is the operating range of the surface agitation boards. The boards, which are usually towed from the spray arms, are troublesome to deploy from some vessels and crews are often unwilling to use them with the result that the effectiveness of the dispersant is greatly reduced. Vessel design often makes it necessary to mount the spray booms midships or aft and with some vessels the bow wave created when travelling at speeds greater than about 5 knots may push much of the oil beyond the dispersants spray swath. This decreases the oil treatment rate and results in wastage of dispersant.

Bow wave effect

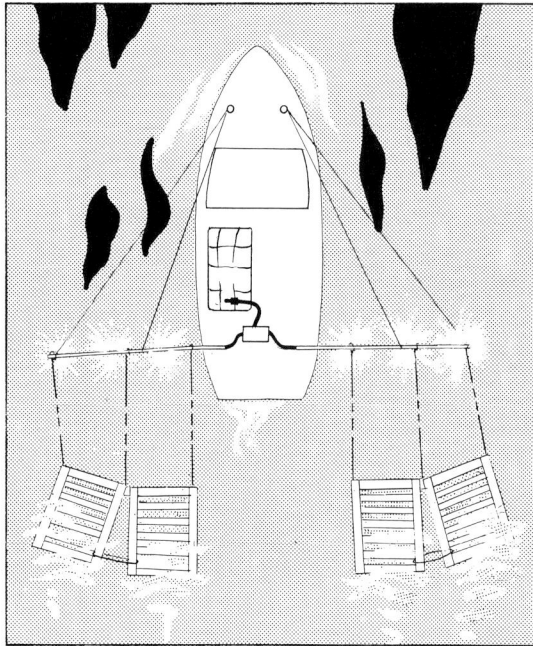

Figure 3a:
Vessel fitted with spray gear for the application of diluted concentrate or hydrocarbon dispersants. Dispersant from a "pillow" tank is pumped through twin booms fitted with spray nozzles and trailing surface agitation boards.

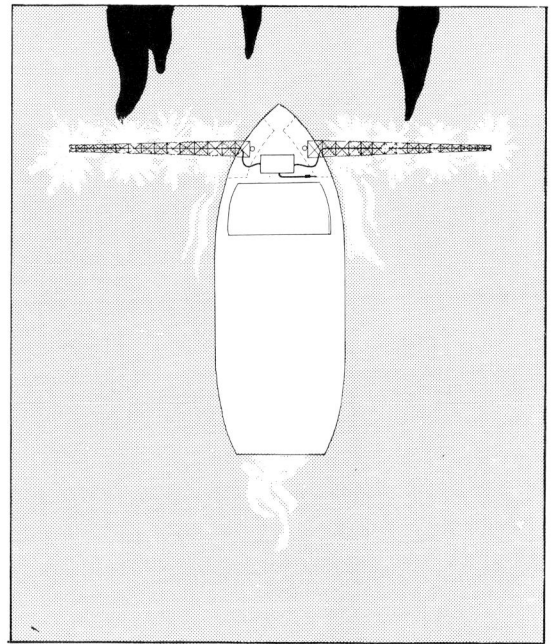

Figure 3b:
Bow-mounted spray gear for the application of undiluted concentrate dispersant. In this example dispersant is drawn from the vessel's storage tanks.

Neat concentrate spraying

Since concentrate dispersants are more effective when used undiluted and require less mixing energy, they allow greater flexibility and higher oil treatment rates. The spray booms can be mounted at the bow which overcomes the problem caused by the bow wave and allows vessels to travel faster while spraying (Figure 3b). In addition, the bow wave itself provides extra agitation. Because the freeboard of most vessels is greater at the bow, the spray booms can be made longer, giving a greater oil encounter width and further improving the potential treatment rate.

Aerial Spraying

Advantages

Despite improvements in vessels' spraying equipment, the technique will always have serious limitations, particularly due to the low treatment rate and the inherent difficulties of locating oil slicks from a vessel. In contrast, aerial spraying offers the advantages of rapid response, good surveillance, high treatment rates, optimum use of dispersant and better evaluation of dispersant treatment.

Undiluted concentrates

Only concentrate dispersants applied neat are suitable for aerial spraying, since they require no additional mixing beyond that provided by wave action. They also make the best use of the available payload.

Weather limitations

Naturally, good visibility over the sea is essential and, since the technique relies upon wave action to break up the slick into droplets, it is not effective in calm conditions.

Aircraft Types

Operating requirements

Helicopters, single and multi-engined fixed-wing aircraft, have all been used successfully for the aerial spraying of dispersant. For an aircraft to be suitable, it should be capable of operating at low altitude and relatively low speeds, within the range 50-150 knots, exhibit good manoeuvrability and carry the highest possible payload for a particular aircraft size.

Crop sprayers

Multi-engined aircraft

Helicopter spray buckets

Aircraft used for spraying fall into three categories: those designed for agricultural or pest control spraying operations, those which have been converted specifically for spraying operations and other aircraft which can be modified at short notice. Aircraft of the first type tend to be single-engined machines with small payloads such as the Piper Pawnee, the Cessna Agtruck and the Bell 47 helicopter. The second group includes multi-engined aircraft, which vary in size from the Piper Aztec to the Douglas DC6. The third group includes most helicopters and large transport aircraft such as the C130 Hercules. Table 3 lists the main characteristics of some typical aircraft suitable for aerial spraying. In addition, many helicopters are equipped with a cargo hook for lifting underslung loads and can therefore carry a 'bucket' spray system without the need for modification.

In order to reduce the high costs of maintaining a large dedicated aircraft, palletised spraying systems have been developed which can be quickly installed in a cargo aircraft.

Selection of Aircraft

The optimum aircraft for a particular operation will be determined primarily by the size and location of the spill, although in reality, aircraft availability will be the crucial factor. In the event of a major spill, payload considerations alone could dictate the use of large multi-engined aircraft. For example, a spill of 4000 tonnes of crude would theoretically require the application of about 200 tonnes of dispersant within 24 hours (12 hours daylight). Taking into account payload considerations only, this could be achieved using one DC6, 10 Piper Aztecs or 20 Piper Pawnees. In reality, many other factors would have to be taken into account, not least of which is the behaviour of the oil itself.

Treatment rates

Location of spill

Small fixed-wing aircraft with high endurance, low fuel consumption, rapid turn-round times and an ability to operate from short and even improvised landing strips are often the most suitable for small spills or fragmented slicks close to shore. However, single-engined aircraft will always be restricted by the distance they can operate safely from the coast. The ability of helicopters to spray in confined situations and operate from a base very near to the spill site is often valuable. The possibility of using them in other roles (e.g. rapid transport of personnel and equipment to inaccessible shore clean-up locations) can be an added advantage. Although larger multi-engined aircraft offer the required range, payload, speed and safety for the treatment of large spills far offshore, it should be recognised that they need longer runways and greater operational support,

TABLE 3. Characteristics of Typical Aircraft suitable for Aerial Spraying of Dispersant

Aircraft Type	Propulsion	Dispersant tank capacity (litres)	Transit speed (knots)	Minimum runway length (metres)
Purpose-built single-engined agricultural aircraft				
Aerospace Fletcher Cresco	turbine	1530	140	300
Aerospace Fletcher	piston	1045	115	245
Antanov An 2 R	piston	1400	100	150
Basant	piston	900	100	215
Cessna Agtruck	piston	1060	100	400
Desmond Norman Fieldmaster	turbine	2640	145	175
EBM 701 Ipanema	piston	680	105	465
IAR-822	piston	600	80	300
Pilatus Porter PC-6	turbine	950	110	180
Piper Brave 300	piston	850	125	295
Piper Pawnee D	piston	570	90	245
PZL Dromader M18	piston	2500	100	250
PZL 106A Kruk	piston	1400	90	220
Super AgCat B	piston	1135	100	180
Thrush Commander	piston	1365	100	300
Turbo Thrush	turbine	2275	125	250
Transavia Air Truk	piston	820	95	335
Converted single & multi-engined aircraft				
Helicopters (fuselage mounted)				
Aérospatiale Lama	1 turbine	1140	80	—
Aérospatiale AS 350	1 turbine	1100	120	—
Bell 47	1 piston	400	75	—
Bell 206	1 turbine	680	115	—
Bell 212	2 turbine	1515	125	—
Hiller UH-12E	1 piston	500	80	—
Hughes 500	1 turbine	680	115	—
Enstrom F-28C	1 piston	400	70	—
Fixed wing				
Beech Baron	2 piston	450	200	410
BN Islander	2 piston	480	140	170
BN Trislander	3 piston	1250	145	395
Canadair CL 215	2 piston	5300	160	915
DC3	2 piston	4600	130	1000
DC4	4 piston	9460	190	1525
DC6	4 piston	13250	210	1525
Fokker F-27	2 turbine	3780	260	990
Grumman Avenger	1 piston	2000	200	915
Lockheed C130	4 turbine	20820	300	1575
Piper Aztec	2 piston	570	175	300
Shorts Sky Van	2 turbine	1200	170	510
Twin Otter	2 turbine	2100	170	320
Volpar Turbo Beech 18	2 turbine	1100	220	510

have longer turn-round times and more restricted visibility and manoeuvrability. Sufficient quantities of the appropriate grade of fuel may be difficult or even impossible to obtain in some parts of the world. This is true for aviation gasoline (Avgas) used to fuel piston engines and in these areas the choice of aircraft is limited to those fitted with turbine engines, which use aviation turbine fuel (Avtur).

Aerial Spraying Equipment

Spray system components

Figure 4 shows the typical components of a system for a small fixed-wing aircraft. A wind-driven pump draws chemical at a controlled rate from one or more tanks to feed the spray booms which are usually fitted close to the trailing edge or above the wing. The chemical is discharged through nozzles or atomisers spaced at intervals along the boom which generate droplets within the required size range.

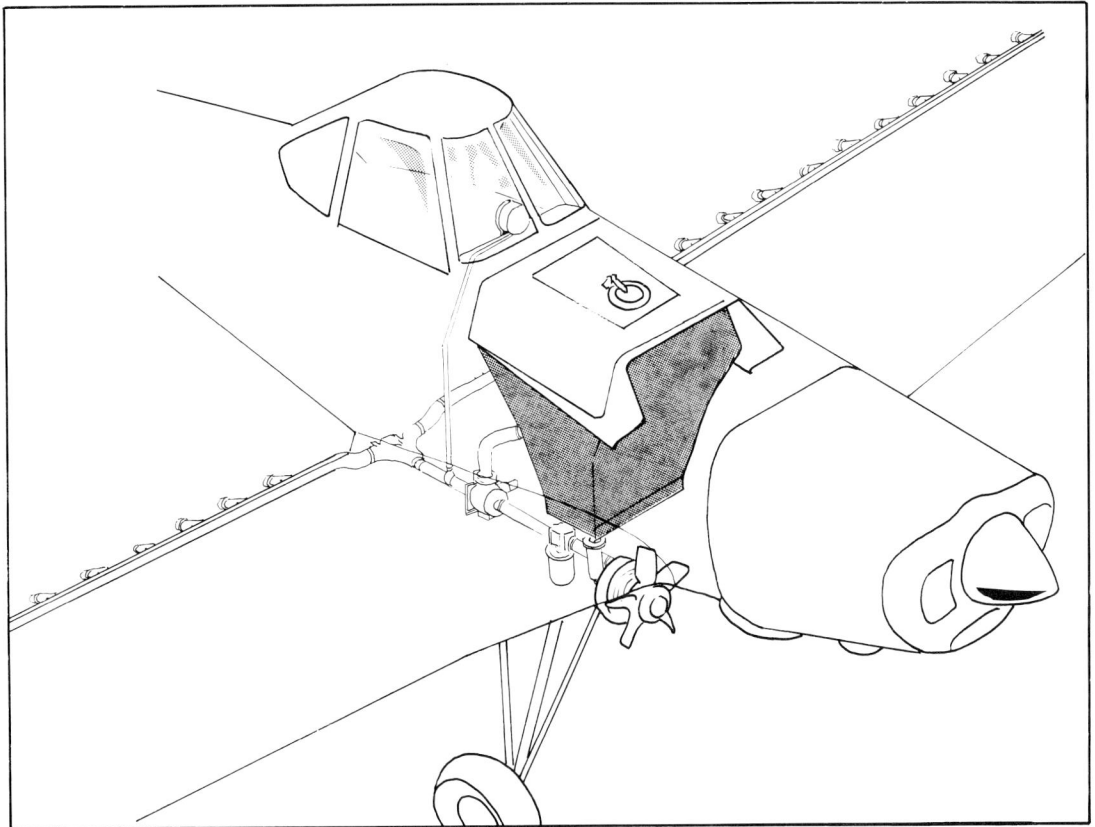

Figure 4:
Spray assembly fitted to a small fixed-wing aircraft. A wind-driven pump under the fuselage feeds chemical from the storage tank to spray booms on the trailing edge of the wings. A gauge fitted in the cockpit allows the pilot to control the flow rate via a by-pass valve.

Spray buckets

Spraying equipment on helicopters is often similar to that fitted to fixed-wing aircraft with the tank and spray boom attached to the fuselage. Alternatively, they may carry a combined tank, pump and spray boom assembly suspended from the cargo hook by wire strops with a quick coupling mechanism (Figure 5, overleaf). The spraying is controlled directly from the cockpit. By using two such 'bucket' spray units alternately, turn-round times can be reduced. Whilst these also have the advantage that the spray pattern is less likely to be affected by the rotor down draught, it is more difficult to maintain the necessary low altitude when using an under-slung system unless a short strop length is used. However, the strop must be at least 1.5 m (5 ft) long to allow sufficient room for the 'bucket' to be disconnected in safety.

Figure 5:
Underslung spray bucket consisting of a dispersant storage tank, pump and spray boom.

The devices used to control drop size fall into two main categories (Figure 6).

Spraying devices

Pressure nozzles (e.g. Spraying Systems 'Tee-Jet', Delavan 'Raindrop' nozzle) which are fitted at intervals along a spray boom.

Rotary devices (e.g. Micronair Rotary Atomiser) which consist of a wind-driven rotating cylindrical gauze cage through which the chemical is pumped forming droplets of the required size range. These are more widely spaced and mounted either on a spray boom or special brackets.

Both pressure nozzles and rotary atomisers are used on helicopters, single-engined crop-spraying aircraft and on the smaller twin-engined aircraft. Pressure nozzles are also used on larger aircraft such as the Douglas DC4 and DC6 because the rotary devices are not considered suitable for these aircraft due to the large number of units that would be required to give the necessary throughput.

Dispersant droplet size

Although dispersant spraying is in many ways similar to agricultural chemical spraying, it differs in two important respects: application rates tend to be higher (typically 100 litres/hectare, 10 imp. gal/acre) and drop sizes larger (500 micrometres (μm) average diameter), the latter to minimise wind drift and possible evaporative losses.

'Tee Jet' nozzles

Apart from the Delavan 'Raindrop' nozzles, the above devices were originally designed to produce the small droplets required for agricultural spraying. However, they can be quickly and easily modified to produce larger droplets. In the case of 'Tee-Jet' nozzles, the swirl plate should be removed and an orifice plate selected to give an opening of about 5 mm. In addition, the nozzles should be oriented parallel to the airflow and pointed aft.

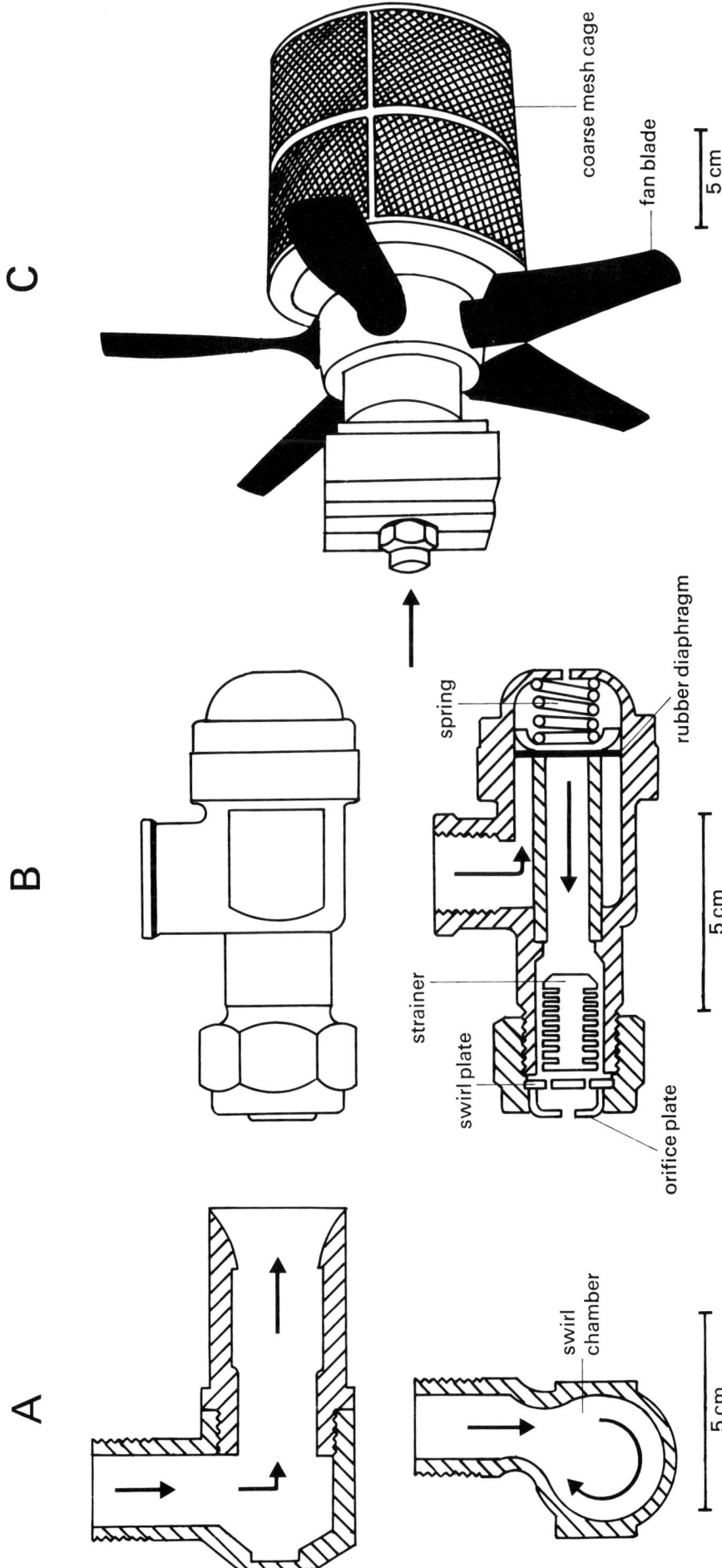

Figure 6:

Spray nozzles. Simplified diagram showing operating principles and directions of dispersant flow.
A. Delavan "Raindrop" nozzle type RA featuring tangential entry of fluid to swirl chamber. **B.** Spraying Systems "Tee-Jet" nozzle. To obtain large droplets the strainer, swirl plate and on larger aircraft also the orifice plate should be removed. **C.** Micronair Rotary Atomizer fitted with variable pitch fan blades and 5-inch cage containing fluid distributor. To obtain large droplets, a smaller 3½-inch cage should be substituted.

III.13

'Micronair' units

The rubber diaphragms fitted to the 'Tee-Jet' nozzles to prevent loss of dispersant when the pump is switched off deteriorate rapidly in contact with dispersant and need to be inspected and replaced frequently. Alternative materials, resistant to dispersant are available. Modification of the Micronair units involves the installation of a 3½" cage of 10 mesh coarse gauze. It is also necessary to slow down the rotation speed by feathering the fan blades to about 70 degrees for the standard 11½" diameter propeller.

Spraying equipment is prone to blockage, particularly with some dispersants which can form a gel when contaminated with small quantities of water. All items of spraying equipment should therefore be regularly inspected and properly maintained. Operating crews should be given comprehensive training in its installation and methods of use and practical exercises should be held frequently.

Application Rates

The application rate required is determined by the type of oil, its thickness and the prevailing conditions. Control of the application rate can be achieved in two ways, either by varying the pump discharge rate or by varying the vessel or aircraft speed while holding the discharge rate constant. The general relationship between the variables is shown below:

pump discharge rate = 0.003 x application rate x speed x swath
   (litres/min)   (litres/hectare)   (knots)   (m)

Example calculation

For a slick estimated to be about 0.2 mm thick, which represents an oil volume of 2 m³/hectare, an application rate of 100 litres/hectare (10 imp.gal/acre) would be required if a concentrate dispersant was used at a dosage of 1:20. A vessel travelling at 10 knots with a swath width of 30 metres would need to discharge dispersant at a rate of 90 litres/min (20 imp. gal/min) to achieve this application rate. If an aircraft flying at 90 knots with an effective swath width of 15 metres was used to treat the same slick, a discharge rate of 405 litres/min (90 imp. gal/min) would be required.

Thicker patches

Flow meters

It is important to concentrate on the treatment of the thickest regions of the slick since application of dispersant to sheen will result in wasted chemical. Whilst application rates of the order of 100 litres/hectare have been found to be appropriate in many situations, thicker patches may call for an increase of the application rate or multiple application in order to achieve the required dosage. Discharge rates can be varied providing the nozzles selected produce a stable spray and the discharge is monitored by a flow meter. In very cold conditions, attention should be given to the viscosity of concentrate dispersants which limits the performance of both nozzles and flow meters.

Vessel spraying

Equipment currently available for applying neat concentrate from ships provides swath widths of between 30 and 40 metres with flow rates ranging from 36 to 220 litres/min (8 to 48 imp. gal/min). Such high capacity spray systems have the potential for excessive application rates leading to wastage of dispersant and it may be necessary to interchange nozzles in order to achieve the optimum application rate.

Special Considerations for Aerial Application

Wind drift

'Windrows

It is essential to minimise dispersant losses due to wind drift and air turbulence. Large droplets assist in this respect but in addition, the aircraft should be flown as low as safety considerations allow. Typically an altitude of 5-15 m (16-50 feet) is used depending on the size of aircraft and the prevailing conditions. Because of the difficulty of judging height over the sea in the absence of a familiar reference point, some aircraft, particularly the larger ones, are equipped with radio altimeters to assist the pilot to fly at these very low altitudes. Wind drift can be further limited by flying into the wind whilst spraying. This is usually necessary anyway due to the tendency for floating oil to become aligned in the direction of the wind as narrow bands or 'windrows' interspersed with zones of thin sheen or clean sea.

Calibration tests

In order to assess dispersant losses and so establish the effective application rate for a particular combination of aircraft and spray system, it is advisable at the planning stage, to carry out calibration tests overland, simulating actual spray sorties. The effect on application rate of such factors as aircraft speed, height, wind speed and direction, swath, dispersant droplet size and dispersant flow rate should be examined using dispersant to which a dye has been added. Such tests are conveniently carried out on an airfield where sampling cards are placed across the expected swath so that the application rate and droplet size distribution can be analysed. This procedure also provides an opportunity to optimise the position and spacing of nozzles for the most uniform pattern across the swath, compensating for such effects as propeller wash and wing-tip vortices.

Control of Spraying Operations

Controlling aircraft

Whichever method is employed to apply dispersants at sea, an objective and continuous assessment must be made of their effectiveness to prevent wastage of costly chemical. To ensure that a vessel or aircraft spraying operation is conducted effectively it is recommended to control it from an aircraft overhead. This can be a light aircraft or helicopter but it must have a high endurance and good communications both with the spray aircraft or vessel and the ground control centre.

Aircraft role

The controlling aircraft can be used in a number of roles. It can be used to identify the heaviest concentrations of oil or those slicks posing the greatest threat, to direct spray craft on to the target, and to judge the accuracy of the application and the effectiveness of the treatment. These functions are particularly important for vessel spraying operations when the range of vision is limited. Air support is also recommended when large multi-engined aircraft are used for spraying because once these aircraft are at low altitude, the crew have great difficulty in distinguishing between oil and sheen, especially if the slick is broken up. When using small spraying aircraft, all these tasks can be undertaken by the pilot, provided he is experienced in the technique.

Organisation and Safety

Logistics

Good organisation is needed to maintain spraying operations for the maximum available time during daylight hours. This may require routine maintenance of equipment and replenishment of supplies during the night. Supplies of fuel and dispersant for aircraft and vessels must be maintained. While dispersants in drums will be adequate to support a small-scale operation, a larger one would have to be supplied from road tankers using high capacity pumps.

Safety procedures

In the case of aerial spraying, relief crews may be needed as flying over the sea at very low altitude is extremely arduous. All the usual safety procedures must be observed despite difficult conditions. Personnel operating on vessels and employed in support activities should be aware of precautions to be observed when using dispersant and should consult the relevant manufacturer for specific guidance.

Deleterious effects

Dispersants create a slippery and therefore unsafe work area especially on the deck of a vessel and spills should be minimised by the promotion of good housekeeping. Dispersants also exhibit a strong degreasing action and regular checks are recommended to ensure that they do not contaminate lubricants, especially in the tail rotor assembly of helicopters, or attack exposed rubber components. They tend to have a deteriorating effect on many paint coatings and may cause slight crazing of stressed Perspex used in windscreens and windows. Vessels and aircraft should therefore be hosed down regularly using fresh water, which for the latter, is to remove salt water spray as well as dispersant.

DISPERSANT SPRAYING ON SHORELINES

Natural clean-up

Dispersants can also be used on some shorelines, particularly during the final stages of clean-up. However, when pollution is heavy, it is important to remove the bulk of the stranded oil first by other means. Beaches subjected to strong wave action are often cleaned naturally and they should be left alone unless circumstances dictate the immediate removal of all the oil.

Application rates

Both hydrocarbon-based dispersants and concentrates may be used for shore cleaning although the former type may be more effective with viscous oils because of the greater penetration achieved by the hydrocarbon solvent. Dispersant application rates should be in the same range as for use at sea. However, it is difficult to predict how effective dispersants will be against stranded oil and it is advisable to carry out a small-scale field test before mounting a large-scale operation. In the case of severe contamination, it is often preferable to clean oiled surfaces with two or more separate treatments, rather than trying to remove all the oil with one application.

Tidal flushing

When using dispersants on shorelines, it is important to apply them in such a way that the beach is washed down with sea water within about 30 minutes to minimise penetration into the beach material. For shingle beaches a shorter period is advisable. The recommended way of using dispersant on tidal shorelines is to spray the oil just ahead of the advancing flood tide. On non-tidal shores gentle hosing with salt water can be considered, taking care not to drive the oil down into the substrate.

Spraying equipment

The most appropriate equipment for application depends on the type of beach material to be cleaned, the ease of access and the scale of the operation. For small inaccessible beaches, portable back-pack sprayers, are the most suitable. For large expanses of beach, purpose-built vehicles, tractors or aircraft can be used.

Rock cleaning

Dispersants can also be used to clean rocks, sea walls and other man-made structures, although it is often necessary to use hand brushes to aid mixing, and high pressure washing to remove treated oil from vertical surfaces and the undersides of rocks. Gelling agents mixed with the dispersant as it is applied can be used to prevent the dispersant running off vertical surfaces so that contact time between the oil and dispersant can be increased.

Viscous oils

Viscous oils, such as heavy fuel oil, can sometimes be dislodged from rocks and sea walls by using the dispersant as a releasing agent. Very little of the oil subsequently washed off will actually disperse and it must therefore be collected with sorbents or skimmers.

Monitoring

Monitoring the effectiveness of beach spraying operations can usually be carried out from the shoreline. It is important to ensure that the treated oil remains dispersed and does not re-surface to form a slick which could contaminate other areas.

Protective clothing

It is recommended that protective clothing is worn by those involved in spraying operations which should include gloves, boots and goggles. While spraying is in progress, access to the public should be restricted.

ENVIRONMENTAL CONSIDERATIONS

Toxicity data

Although the toxicity of modern dispersants is far less than earlier formulations, their use still evokes much controversy. Different countries have different attitudes toward the use of dispersants. Concern stems mainly from the fact that the use of dispersants represents not only a deliberate introduction into the sea of an additional pollutant but can also result in a local increase in hydrocarbon concentration in the water column, which may lead to biological damage. There is a wealth of laboratory data on the toxicity of dispersants and oil/dispersant mixtures to a variety of marine organisms, but far less is known about their effects at actual spills where dilution usually reduces concentrations and exposure times significantly. On the basis of field trials, it has been found that initial concentrations can reach 50 mg/litre but fall rapidly to background levels within a few hours.

Field studies

The few studies undertaken in areas where dispersants have been used extensively have not demonstrated significant effects on populations of particular species or biological communities. Conclusive evidence of increased tainting of commercial species resulting from dispersant usage is also not available.

Dilution potential

An assessment of the dilution potential is the most useful basis for deciding whether dispersants can be used for protecting certain resources without risking undue damage to others. Relevant factors are the distance between the application site and sensitive areas as well as the direction of currents and the mixing depths of surface waters. It is frequently possible to make a rough estimate of both the maximum likely dispersant/oil concentration and the exposure time for a specific location. In areas where the dilution potential is great, such as the open sea, elevated concentrations are unlikely to persist for more than a few hours and significant biological effects are therefore improbable.

Water exchange

In shallow waters close to the shore where water exchange is poor, higher concentrations may persist for long periods and possibly approach values known to cause observable effects in laboratory experiments. Because of this, great concern is often expressed about the use of dispersants in such areas. Despite this greater risk, the careful application of dispersants may on these occasions be beneficial overall, if damage to adjacent ecologically sensitive shorelines is reduced as a result.

Balancing priorities

The circumstances favouring the use of dispersants are seldom clear-cut and the choice is necessarily a compromise between other options, cost-effectiveness and conflicting priorities for protecting different resources from pollution damage. On occasions the potential benefit gained by using dispersants to protect coastal amenities, sea birds and intertidal marine life may far outweigh any potential disadvantages, such as the temporary tainting of commercial shellfish. Conversely, the dispersant option may be rejected in fish spawning areas in the open sea even if the risk of damage is very low and would not normally be advised close to industrial water intakes, shellfish beds, coral reefs or within wetland regions. Despite the difficulties, it is important that an order of priority for the resources to be protected is established and the circumstances under which dispersants may be used are agreed upon before the occurrence of a spill.

PRE-SPILL PLANNING

Sensitivity maps

Before an incident occurs, it is important to identify and map all areas sensitive to dispersant and dispersed oil such as fish spawning areas, shellfish beds, wetland areas, tidal flats and lagoons. The case for using dispersants in different locations and circumstances should then be examined in advance so that equipment and materials can be sited and stored accordingly. Such measures reduce the response time and improve the chances of a successful operation.

Laboratory and Field Tests

Effective-ness

Toxicity

In many countries testing and approval procedures have been developed to ensure that approved dispersants are effective, biodegrade rapidly and have minimal toxicity to marine life when used correctly. The effectiveness of different dispersant chemicals varies considerably and it is always advisable to carry out either laboratory or field tests to identify the best available product under the conditions expected. Environmental factors, such as water temperature and salinity, and the probable types of oil spilled should be considered. Similarly, by conducting toxicity tests using locally available marine organisms, low toxicity formulations can be selected and a list of approved dispersants developed.

Stockpiling and Storage

Quantity stockpiled

Bulk storage

When a dispersant is being stockpiled for use in time of an emergency, careful consideration has to be given to the quantities needed in relation to the most likely size of spill expected. Consideration must also be given to the logistics of manufacturing and transporting additional supplies in the event of a major spill. Dispersants can either be stored in 200 litre drums or in bulk containers. In this latter case, precautions should be taken to prevent contamination from previously stored materials and from the ingress of water. If dispersants of different manufacture are to be stored together in bulk, it is important to ensure that they are compatible; hydrocarbon solvent-based dispersants and concentrates should not be mixed since this could lead to the formation of viscous gels.

Drums

Vessel storage

For quick response, storage in road tanker trailers can be particularly useful. Although most dispersants have a long shelf life, steel drums can often corrode from the outside leading to loss of chemical. Plastic drums are therefore preferable for long-term storage provided they are kept out of direct sunlight. Whether stored in bulk or in drums, portable transfer pumps will usually be necessary to reduce delays when loading vessels or aircraft. Where dedicated vessels are employed it may be possible to store the chemical in integral tanks on board. For most other vessels, collapsible "pillow" tanks, of up to 4500 litre capacity (1000 gallons), can conveniently be placed on deck.

Provision of Vessels and Aircraft

Location of materials and equipment

For dispersant application from vessels, the location and availability of tugs, fishing boats and other vessels should be ascertained if it is not planned to have dedicated vessels on permanent stand-by. If aerial application is envisaged, landing sites should be identified along the coast and consideration given to aircraft maintenance, sources of supply for dispersant and fuel, and arrangements for transporting them.

Agri-cultural aircraft

Multi-role aircraft

Although locally available agricultural aircraft can be utilised for small spills these may be difficult to obtain at certain times of the year. Major spills, particularly those further offshore, require larger aircraft and since these are usually not readily available it may be necessary to provide a dedicated response capability. Since this would involve the payment of a substantial retainer it is advantageous to find a compatible additional role for the aircraft. One such role might be firefighting which has many parallels with dispersant spraying. Alternatively, they might be used to monitor traffic separation schemes and to detect and identify vessels discharging oil illegally into the sea. Crew familiarity with both low flying over the sea and recognition of oil could be maintained in this way.

POINTS TO REMEMBER

1. The application of dispersant enhances the natural break-up of oil thereby removing it from the water surface and protecting shorelines and natural resources such as sea birds.

2. Two types of dispersant are available, hydrocarbon-based and concentrates. Concentrates may be used neat or diluted to 10% with seawater but are more effective undiluted. Typical dose rates are between 1:5 and 1:30 neat dispersant to oil. Hydrocarbon solvent-based dispersants must not be diluted with water and typical dose rates are between 1:1 and 1:3 dispersant to oil.

3. It is important to recognise the limitations of dispersants, in particular their inability to treat viscous oils and water-in-oil emulsions. Since most crude oils spilled at sea rapidly become resistant to dispersants, a fast response and high treatment rate are essential.

4. Whilst vessels are suitable for dealing with small spills close to port, aircraft potentially offer a more cost-effective response for larger spills offshore.

5. Aerial spraying has the advantages of rapid response, good surveillance, high treatment rates, optimum use of dispersant, and good evaluation of dispersant treatment.

6. Dispersants should only be used for beach cleaning after gross pollution has been removed. Care must be taken to prevent the penetration of oil into beach material and to reduce damage to marine life caused by prolonged exposure to dispersant/oil mixtures.

7. Success depends upon positive control of spraying operations which demands a comprehensive communications network.

8. In order to formulate a policy on the use of dispersants, the risk of environmental damage resulting from their use must be balanced against the probable effects of the untreated oil. Dilution potential will be an important consideration. The policy for dispersant use should be agreed in advance for each area.

FURTHER READING

CONCAWE (1981) A field guide to coastal oil spill control and clean-up techniques. CONCAWE, The Hague, Netherlands, Report No. 9/81. 112 pp.

Cormack, D. (1983) Response to oil and chemical marine pollution. Applied Science Publishers, London. 531 pp.

Croquette, J. (1985) The use of dispersants at sea to control oil slicks. CEDRE, Brest, France. Publication No. R85.70.E. 35 pp.

IMO (1980) Manual on oil pollution — Section IV Practical information on means of dealing with oil spillages. IMO, London. 143 pp. (under revision).

IMO/UNEP (1982) Guidelines on oil spill dispersant application and environmental considerations. IMO, London. 43 pp.

Institute of Petroleum (1986) Guidelines on the use of oil spill dispersants (Second edition). John Wiley & Sons, London.

ITOPF (1982) Aerial application of oil spill dispersants. Technical Information Paper No. 3, ITOPF, London. 8 pp.

ITOPF (1982) Use of oil spill dispersants. Technical Information Paper No. 4, ITOPF, London. 8 pp.

Morris, P.R. and Martinelli, F. (1983) A specification for oil spill dispersants. Warren Spring Laboratory, Stevenage, U.K. Report LR 448 (OP) M. 21 pp.

IV SHORELINE CLEAN-UP

Many oil spills result in pollution of shorelines despite efforts to combat the oil at sea and to protect the coastline. Shoreline clean-up is usually straightforward and does not normally require specialised equipment. However, the use of inappropriate techniques and inadequate organisation can aggravate the damage caused by the oil itself.

This section describes techniques that have been applied successfully on a number of commonly occurring shore types. Shoreline clean-up usually results in the collection of substantial quantities of oil and oily debris for which various disposal options are outlined.

CONTENTS

CLEAN-UP STRATEGY

Assessment of the Problem

Evaluation

Before taking any action to clean up oil on shorelines, it is necessary to determine the type and amount of oil involved, the geographical extent of the pollution, and the length and nature of the affected coastline. In addition, it is important to identify the source of the oil, so that the likelihood of further shoreline impact can be evaluated, and to allow subsequent claims for compensation for damages and clean-up costs to be directed to the appropriate party.

Estimating oil quantity

Estimating the amount of stranded oil with accuracy is difficult on most types of shorelines because the distribution of oil is seldom uniform. On exposed rocky shores the task can be particularly hard due to the numerous holes and crevices where oil may collect. However, even a rough estimate of oil quantity is desirable for the purpose of organising the most appropriate shore clean-up response and identifying the manpower requirements for the task.

Shoreline survey

The overall extent of pollution can be assessed visually by first overflying the area. A more detailed evaluation of the oil present on a short representative section of the affected shorelines can then be made on foot. This has to be repeated on other sections where the degree of oil coverage may be different or where the character of the shoreline changes (Table 1). At the same time, the shoreline survey provides a good opportunity for securing oil samples in case chemical analysis is required, and for confirming access routes and the feasibility of clean-up.

On occasions, oil on shorelines will be best left to weather and degrade naturally. When shoreline clean-up is required, three stages can usually be recognised:

Stage I Removal of heavy contamination and floating oil;

Stage II Clean-up of moderate contamination, stranded oil and oiled beach materials;

Stage III Clean-up of lightly contaminated shorelines and removal of oil stains.

Priorities

Priorities for action will need to be decided after consideration of potential conflicts of interest. For example, the use of the most effective techniques may be damaging to some environmentally sensitive habitats, whilst elsewhere amenity interests may overrule such considerations. This will require a balanced judgement on a site by site basis.

Stage I

At the first stage, floating oil reaching the coast should be contained and collected as quickly as possible to prevent it from moving to previously uncontaminated parts. This also applies to heavy concentrations of stranded oil which may otherwise float off, particularly in tidal regions. Booms can sometimes be used to hold the oil against the shore while the operation is in progress. This strategy may, however, be inappropriate for environmentally sensitive shorelines where it may well be better to allow the oil to migrate to less sensitive areas.

Stage II

If there is no risk of oil migration, it is usually preferable to wait until all the oil from a particular incident has come ashore before beginning beach clean-up to avoid cleaning the same area more than once. However, this must be weighed against the likelihood of the oil becoming mixed into the substrate and even buried if the clean-up is delayed too long. This stage of shoreline clean-up is often the lengthiest. Great care is needed to limit the quantities of beach material removed with the oil so that both the risk of subsequent erosion and the quantity of material for disposal are minimised.

Table 1: BEHAVIOUR OF OIL ON SOME COMMON TYPES OF SHORELINE

	Type & Size Range	Comments
	Rocks, Boulders & Man-made Structures >250 mm	Oil is often carried past rocky outcrops and cliffs by reflected waves but may be thrown up onto the splash zone where it may accumulate on rough or porous surfaces. In tidal regions, oil collects in rock pools and may coat rocks throughout the tidal range. This oil is usually rapidly removed by wave action but is more persistent in sheltered waters.
	Cobbles, Pebbles & Shingle 2–250 mm	Oil penetration increases with increasing stone size. In areas with strong wave action, surface stones are cleaned quickly by abrasion whereas buried oil may persist for some time. Low viscosity oils may be flushed out of the beach by natural water movement.
	Sand 0.1–2 mm	Particle size, water table depth and drainage characteristics determine the oil penetration of sand beaches. Coarse sand beaches tend to shelve more steeply and dry out at low water enabling some degree of penetration to occur particularly with low viscosity oils. Oil is generally concentrated near to the high water mark. Fine grained sand is usually associated with a flatter beach profile remaining wet throughout the tidal cycle so that little penetration takes place. However, some oil can be buried when exposed to surf conditions for example, during a storm.
	Mud (mud flats, marshes, mangroves) <0.1 mm	Extensive deposits of mud are characteristic of low energy environments. Little penetration of the substrate by oil occurs because the sediment is usually waterlogged but oil can persist on the surface over long periods. If the spill coincides with a storm, oil can become incorporated in the sediment and persist indefinitely. Animal burrows and plant root channels can also bring about oil penetration.

Clean-up termination

Stage III

It is often difficult to decide at what stage the clean-up operation should stop. It is usually determined by such factors as the importance given to the area, the time of year and the rate at which natural cleansing is expected to take place. Costs also play a part in this decision since the effort required to achieve an improvement will rise disproportionately as the amount of oil remaining on the shore decreases. An exhaustive final clean-up stage is therefore generally only required for high amenity areas during, or just before, the tourist season.

CLEAN-UP TECHNIQUES

Rocks, Boulders & Man-made Structures

Stage I

Mechanical recovery

Where vehicles or boats can reach the water's edge the oil can be collected using skimmers, pumps, or vacuum trucks. Many skimmers do not perform well in shallow water or in waves and a vacuum truck or tank trailer may work better. Typically these can collect 20 m³ per day. As far as possible, free water collected with the oil should be allowed to settle and then be drained off before the oil is taken away for disposal. On some shorelines the oil can be flushed off the rocks or stones and collected with booms or floating ropes.

Manual recovery

Where vehicles are unable to get sufficiently close to the water's edge, the oil has to be picked up manually using buckets, scoops or other containers. Open-topped 200-litre drums are unsuitable because they are hard to manhandle over rocky terrain when full. However, drums can be carried in small boats for collecting oil trapped amongst rocks. If the oil is particularly fluid it may be easier to handle if sorbents are first mixed into it.

Sorbents

The most effective are synthetic materials such as expanded polyurethane foam and polypropylene fibres. These tend to be expensive although some can be used several times. In the absence of synthetic products, naturally occurring local materials such as straw, palm fronds, coconut husks, sugar cane waste (bagasse), peat and chicken feathers can be used. The oil/sorbent mixture can then be collected with forks and rakes and carried from the collection point in heavy gauge plastic bags or small containers. Using too much sorbent can be a nuisance if it causes secondary pollution and a good rule is not to apply more than can be readily recovered.

Stages II & III

High pressure washing

In many cases, once the mobile oil has been removed, the oil remaining on rocks, boulders and man-made structures can be left to weather since a hard surface film readily forms, minimising the spread of pollution. However, where rocky shores are part of the coastal amenities, further cleaning can be achieved by washing with water under pressure. Either hot or cold water can be used depending upon equipment availablility and oil type: higher temperatures and, on occasions, even steam are required to dislodge viscous oils. Typically water is heated to about 60°C and sprayed at 10-20 litres/minute from a hand lance operating at between 80 and 140 bar. Oil released in this way must be collected, otherwise it may pollute previously cleaned or uncontaminated surfaces. The oil may be flushed down into a boom at the water's edge and collected with skimmers or vacuum trucks (Figure 1) or it may be collected by arranging sorbents at the base of the surface being cleaned. In tropical and sub-tropical environments hot water washing is likely to be less effective than in temperate climates since oil exposed to the sun becomes baked on to the rock. Small areas can be cleaned by sand blasting.

Collection of released oil

Figure 1:
Oil flushed off a stone embankment into a boom for recovery by vacuum truck.

Adverse effects

Whereas many marine plants and animals will survive a single oiling, any of the methods described above will lead to the destruction of most of the marine biota. Some physical damage to the treated surfaces themselves may also occur. These methods should therefore be reserved for areas where there is easy access and where members of the public are likely to come into contact with oil if no action is taken.

Dispersants

The use of dispersants can sometimes also assist oil removal although their use should be restricted to areas where water movement will allow rapid dilution and so prevent damage to sensitive marine life. In some instances, and particularly with more viscous oils, the dispersant acts simply to dislodge the oil from the surface and does not produce a dispersion. In such cases every effort should be made to collect this oil to prevent recontamination.

Removal of stains Stains can sometimes be removed by brushing dispersant into the oil or by applying it as a gel and then hosing off the oil/dispersant mixture. By this stage of the clean-up the oil will be in the form of extremely thin films and so only very light applications will be required.

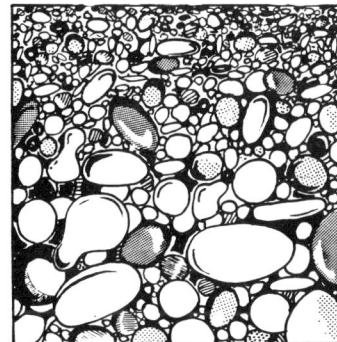

Cobbles, Pebbles and Shingle

Stage I

Oil penetration This type of shoreline is probably the most difficult to clean satisfactorily because much of the oil will have penetrated deep into the beach through the spaces between the stones. The first stage of clean-up for this type of shoreline follows similar lines to the previous one: pumping fluid oil where possible or removing it by hand. The poor load-bearing characteristics of such beaches can hinder the movement of both vehicles and personnel.

Stages II & III

High pressure flushing Water at high pressure can be used to flush surface oil to the water's edge but some of the oil will also be driven into the beach. Low viscosity oils may be washed out from between the stones and the use of dispersants can sometimes enhance this. Inevitably some oil will penetrate further into the beach after the stones at the surface have been cleaned. This will slowly leach out as a sheen over a period of weeks or longer.

Removal of stones The removal of oily stones will rarely be practical and will usually only be possible if tracked front-end loaders can be used. Removal of stones should only be considered if it is certain that it will not cause serious beach erosion and that it will be possible to dispose of the material.

Masking Another approach which might be used in locations subject to vigorous winter storms is to cover the oiled area with stones from higher up the beach, so providing a clean surface during the summer for those using the beach for recreation. Some weathering will occur due to summer temperatures and then, during the natural rearrangement of the beach that takes place in the winter, the oil will be broken up and dispersed. This method can only be used where the beach is moderately oiled and is not suitable for finer beach materials because the oil tends to migrate back to the surface. The beach profile may also be permanently altered and the natural sea defences weakened.

Natural cleaning One way to remove the greasy film that often remains on stones after cleaning is to push the top layer into the sea where the abrasive action caused by the waves rapidly cleans them. However, this is obviously inappropriate if oily stones underneath are then exposed. It should also be appreciated that it may be several years before a cobble beach profile is restored since vigorous wave action is necessary to lift stones of this size back up the beach.

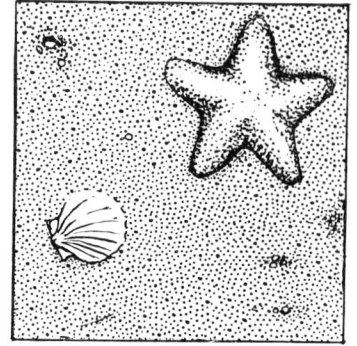

Sand Beaches

Stage I

High priority

Very often sand beaches are regarded as a valuable amenity resource and priority is given to cleaning them. Intertidal sand flats, on the other hand, are often biologically productive and important for commercial fisheries. Environmental considerations may therefore dictate the selection of methods likely to cause the least additional damage, such as those described later for muddy shores.

Mechanical equipment vs. manual methods

Recreational beaches often have good access although on some shorelines temporary roadways may have to be constructed to allow heavy equipment onto the beach. Whilst bulk oil can be removed relatively easily from sand beaches, a desire to clean them quickly can sometimes lead to difficulties. In a major spill a balance has to be struck between the speed with which large quantities of oil can be collected using heavy machinery and the associated increased contamination of beach substrate. To a large extent this is determined by beach type. Coarse sand beaches are frequently unable to support any vehicle without its wheels or tracks sinking into the sand and causing oil to be mixed further into the beach. Worse still, vehicles driven onto the beach may become immobilised once loaded.

Manual methods

Manual methods must be used where there is no access for vehicles, no hard-standing at the top of the beach, or if it is too far for pump or suction hoses to reach the water's edge. Oil, as well as oiled sorbents and debris, can be collected in heavy gauge plastic bags or other containers and carried up the beach, above the high water mark.

Graders & front-end loaders

Flat hard packed beaches may support heavy vehicles such as graders and front-end loaders. The grader's blade is set to skim just below the beach surface and the oil and sand drawn into lines parallel to the shoreline (Figure 2). Ideally, the grader should start from the clean end of the beach and work upwards. This may have to be modified depending on the pattern of contamination, state of the tide and whether the sand is loose or compact. The accumulated oil is then picked up by front-end loaders. The work can be done using front-end loaders alone although the amount of sand picked up will then inevitably be greater. In all cases, care must be exercised to ensure that excessive removal of sand does not result in beach erosion.

Stage II

Moderately contaminated oily sand and debris is best removed from sand beaches by teams of men working in conjunction with front-end loaders, the latter being used solely to transport the collected material to temporary storage sites at the top of the beach. Typically each person collects between 1 and 2 m^3 per day by this method. Front-end loaders and other heavy machinery can be used to pick up the oily sand directly and can remove as much as 100-200 m^3/day/machine. As a rule, however, the oil content of sand collected by machines is only between 1 and 2% whereas that collected manually contains 5-10% oil. This is because heavy equipment tends to mix the oil into the sand and is less selective in what it picks up with the result that at least three times as much sand is removed as compared with manual methods.

Ratio of oil:sand

Methodical collection

To make the most efficient use of each front-end loader, the clean-up teams should collect the oily sand into piles or alternatively fill 200-litre drums placed at intervals along the beach. To prevent oil being spread up the beach the front-end loader should work from the clean side as far as possible. Vehicles equipped with low pressure tyres are generally more suitable than tracked vehicles.

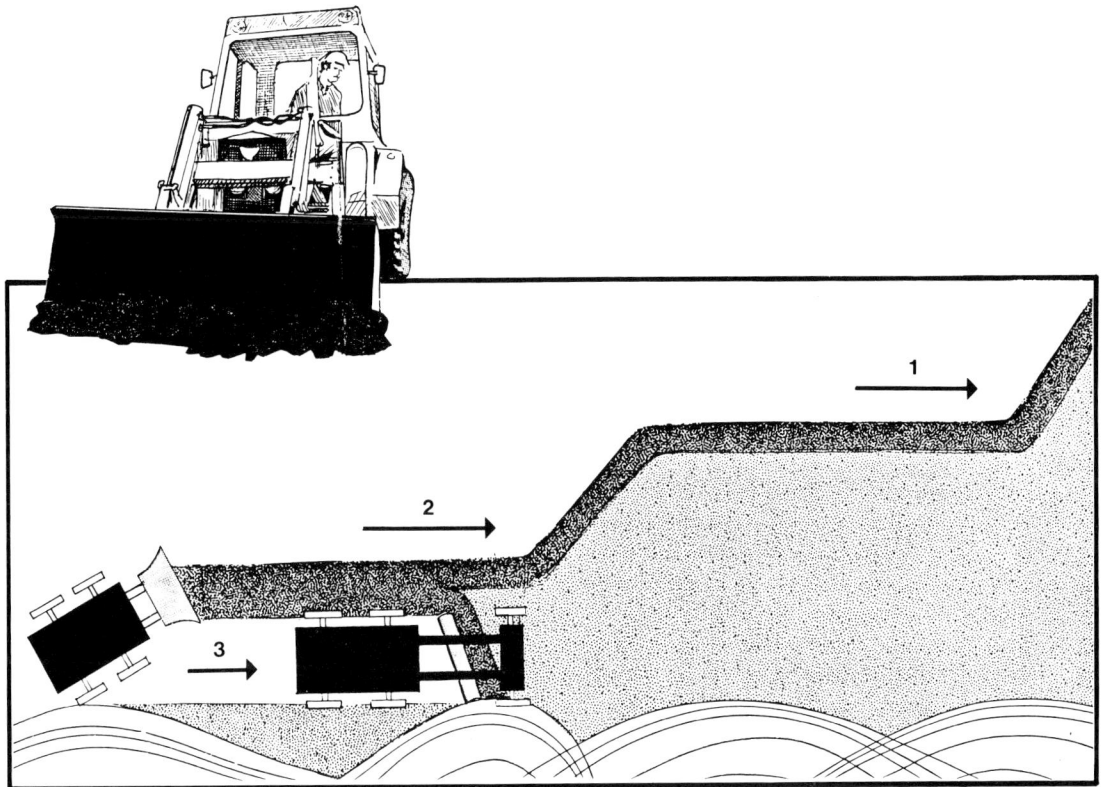

Figure 2:
Grader used systematically on a uniformly polluted beach to concentrate oil into strips for collection by a front-end loader to a nearby temporary storage site.

Use of plastic bags

Where there is no possibility of getting vehicles onto the beach the collected oily sand has to be carried off the beach in plastic bags. Heavy duty bags such as those used for fertiliser are suitable. They should not be filled completely because of the difficulty in carrying them over soft sand. A simple two-man litter can be made to carry the bags (Figure 3). Plastic bags exposed to strong sunlight for more than about 10 days will begin to deteriorate and so disposal of the filled bags should not be delayed.

Figure 3:
Clean-up of moderately contaminated sandy beach by manual methods.

Stage III

Dispersants

After most of the contaminated beach material has been removed, that remaining is likely to be greasy and discoloured. This will not usually be sufficiently clean for recreational beaches and a final stage will be necessary. Dispersants can be used, applied from back packs, agricultural vehicles or aircraft. The dispersant should be allowed about 30 minutes' contact with the oily beach before being washed by the incoming tide. In non-tidal regions or in areas without strong surf action, hosing with sea water may be necessary to achieve a good dispersion.

Ploughing

Another method particuarly appropriate for lightly oiled tidal beaches is periodically to plough or harrow the affected beach at low water. The oil is then mixed with a greater volume of sand and more frequently exposed to weathering processes.

Flooding

Oil can be released from coarse-grained sand by passing high volumes of water through sections of the beach. Sea water is drawn through a high capacity pump and distributed through a number of hoses at low pressure. By directing the water into a small area of beach the oil can be floated out and flushed to the water's edge for collection. The method is slow and limited to the treatment of small areas at a time.

Beach cleaning machines

The material remaining after the clean-up of dry sand beaches is usually in the form of small nodules of oily sand up to about 50 mm in diameter. These, and tar balls washed up along the high water mark, can be picked up using beach cleaning machines which skim the top surface of the beach and pass the sand through a series of vibrating or rotating screens (Figure 4). The oily lumps are retained within the vehicle while the clean sand is allowed to drop back onto the beach. Such machines were designed originally for general beach litter collection.

Figure 4:
Tractor-powered beach cleaning device for collection of tar balls and debris.

Sand replenishment

Prior removal of sand

If the spill coincides with the tourist season it may be necessary to return the beach to its original condition in the shortest possible time. Clean sand can be brought in from elsewhere and spread over any remaining lightly oiled sand. As far as possible, this clean sand should have the same grain size as the natural material so that it does not alter the physical and biological characteristics of the beach. If a finer grained sand were to be used as replacement there is a risk that it might be washed away too quickly. When sufficient notice is available before a spill reaches a beach it may be appropriate to move some of the sand above the high water mark. This material can then be replaced after the beach has been cleaned.

Muddy Shores

Oil damage vs. clean-up damage

Whenever possible it is preferable to allow oil that arrives on this type of shoreline to weather naturally, particularly where it has been washed up onto vegetation. It has often been found that activities intended to clear pollution have resulted in more damage than the oil itself due to physical disturbance and substrate erosion (Figure 5).

Low pressure flushing

Cutting vegetation

Marsh vegetation often survives a single oil coating and in several instances new plants have been found to grow through a covering of oil. Where removal of the oil is essential to prevent its transfer elsewhere, low pressure water hoses can be used to flush it into open water where it may be contained within a boom for subsequent collection. This technique is best applied by approaching the shoreline from the water in shallow draught boats. If birds are threatened, cutting and removal of oiled vegetation might be considered but must be balanced against the longer term damage likely to be caused by trampling.

Mangroves

Similar considerations apply to mangroves. Where the trees are particularly dense and there is a high risk of oil being carried further into mangrove stands, it may be necessary to remove some vegetation to allow access so that the oil can be flushed out. This may prevent destruction of mangroves over a wider area but it should be appreciated that this approach will cause local damage, the recovery of which is likely to be slow.

Figure 5:
Physical damage to salt marsh by vehicle movements.

ORGANISATION

Division of work

Proper organisation of the workforce engaged in shoreline clean-up is vital to the success of the operation. This can be achieved by division of the affected coastline into small areas and it may be appropriate to relate these divisions to shoreline type. A supervisor is assigned to each area with the workforce divided into teams (Figure 6). Each team is allocated a section of the beach to clean based on the amount of material that each person can be expected to collect in a day; for example, on a sandy beach 1-2 m^3/day. The workforce then have the satisfaction of completing a task each day and seeing the progress they have made, while the beach is cleaned methodically section by section.

Small teams

A team should contain about ten people plus a leader, and each supervisor should be responsible for about ten teams. Although in principle military command structures lend themselves well to this type of operation they tend to result in the work-teams being too large so that some modification may be necessary. On tidal shores the work should

Working hours

be arranged to coincide with the tidal cycle with rest periods and meal breaks being taken at high water. Night time working is usually found to be inefficient even when adequate lighting is provided.

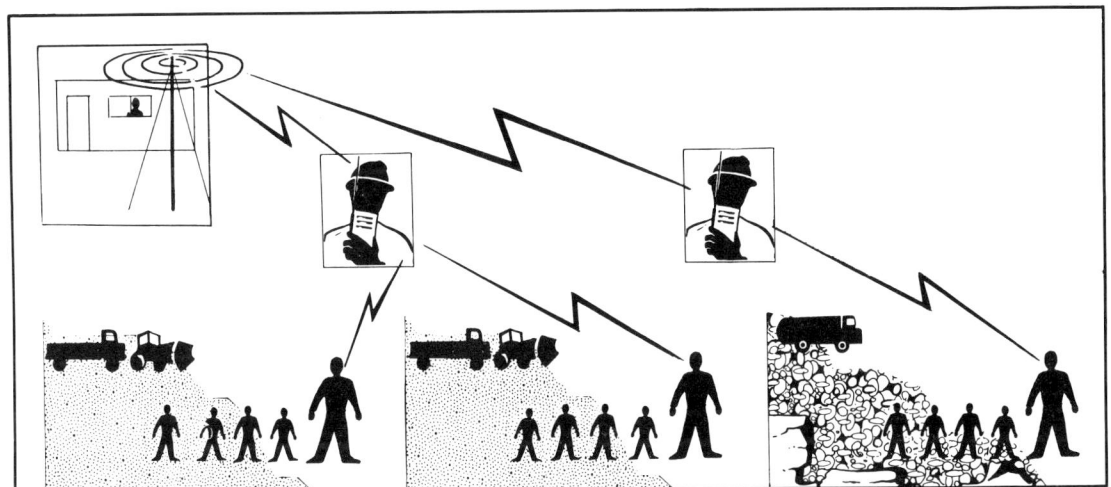

Figure 6:
Schematic diagram of organisation and lines of communications between command post, supervisors and shore clean-up teams.

Control of vehicles

Equipment should be organised to complement the workforce. Vehicles working on the beach should be confined to the work area while larger capacity lorries, transporting the collected material to storage or disposal sites, should be kept off the beach so that dirty and clean areas remain segregated. This not only limits the number of vehicles on the beach but also helps to reduce the amount of oil transferred directly from the beaches to the roads. Spillage onto roads can be further limited by lining the lorries with plastic sheeting.

Traffic control

Access to some work sites will need to be restricted to minimise damage to sand dunes and natural sea defences. Traffic around the work site should be controlled so that lorries are able to move without hindrance. Police may also be required to close the beach and access roads in the interests of public safety, particularly where heavy vehicles are used.

Record keeping

Daily records of the workforce and equipment employed in each area should be kept and are essential for drawing up subsequent claims for compensation. A record maintained at the same time of the quantities of oil and oily debris removed enables progress to be easily monitored. In addition to written reports, the status of each work site and the location of manpower and equipment can be conveniently monitored on large-scale maps.

DISPOSAL OF OIL AND DEBRIS

Ideally, as much of the collected oil as possible should be processed through a refinery or oil recycling plant. Unfortunately this is rarely possible due to weathering of the oil and contamination with debris and so some other form of disposal is usually required. This includes direct dumping; stabilisation for use in land reclamation or construction of minor roads; and destruction through biological processes or burning. The disposal option chosen will depend upon the amount and type of oil and debris, the location of the spill, environmental and legal considerations, and the likely costs involved.

Type and Nature of the Oil and Oily Debris

Fluid oils

As a general rule, only spills of persistent oils such as crude, heavier grades of fuel oil and some lubricants are likely to require treatment and disposal since clean-up of non-persistent oils is not usually necessary. If the oil can be collected soon after being spilled, it is likely to be fluid and relatively free of contamination. In most cases, however, collected oil will be viscous due to weathering. Oil collected from the water will probably be relatively free of solid debris but is likely to contain large amounts of water present as an emulsion. On the other hand, oil stranded on the shore is normally associated with considerable quantities of solids and is often difficult to separate in a form suitable for recycling. Three main types of waste may be collected from the shore: oil mixed with sand; oil mixed with wood, plastic or seaweed; and solid tarballs. Each may require a different method of treatment and disposal (Table 2).

Solid wastes

Storage and Preparation for Disposal

Temporary storage

The large volumes of material requiring disposal following clean-up can often present major problems of handling and transportation. It is usually necessary therefore to store the material temporarily to provide a buffer between collection and final disposal and to allow time to select the appropriate disposal method. In the case of material resulting from shore clean-up, temporary storage at the back of the beach also enables transport to be split into two stages: from beach to temporary storage and then, at some future time, from temporary storage to the final disposal site.

Viscous oils

As far as possible, bulk oil should be stored separately from oily debris so that different methods of treatment and disposal can be followed. Provided the oil is pumpable at ambient temperatures, it can be stored in enclosed tanks. However, care should be exercised in the bulk storage of more viscous materials, particularly if the tanks are not fitted with heating coils, since it may prove difficult to empty them. Highly viscous oils are best stored in open containers such as barges, skips or drums to facilitate treatment and transfer operations.

Storage pits

If special purpose containers are not available, bulk oil from shorelines can often be held within compacted earth walls or in simple storage pits lined with heavy gauge polyethylene (or other suitable oil-proof material). Long narrow storage pits approximately 2 m wide and 1.5 m deep are preferable (Figure 7). If there is a possibility of heavy rainfall, the pits should not be filled too full to ensure that oil does not overflow. Where temporary storage of bulk oil is required in sensitive areas such as sand dunes, it is important to avoid disturbance of the stabilising vegetation since this could lead to erosion. Pits should be filled in after complete removal of the oil and, as far as possible, the area restored to its original state.

Restoration of storage areas

Table 2: OPTIONS FOR SEPARATION & DISPOSAL OF OIL AND DEBRIS

Type of Material	Separation Methods	Disposal Methods
LIQUIDS		
— Non-emulsified oils	— Gravity separation of free water	— Use of recovered oil as fuel or refinery feedstock
— Emulsified oils	— Emulsion broken to release water by: — heat treatment — emulsion breaking chemicals — mixing with sand	— Use of recovered oil as fuel or refinery feedstock — Burning — Return of separated sand to source
SOLIDS		
— Oil mixed with sand	— Collection of liquid oil leaching from sand during temporary storage — Extraction of oil from sand by washing with water or solvent — Removal of solid oils by sieving	— Use of recovered liquid oil as fuel or refinery feedstock — Direct disposal — Stabilisation with inorganic material — Degradation through land farming or composting — Burning
— Oil mixed with cobbles, pebbles or shingle	— Collection of liquid oil leaching from beach material during temporary storage — Extraction of oil from beach material by washing with water or solvent	— Direct disposal — Burning
— Oil mixed with wood, plastics, seaweed and sorbents	— Collection of liquid oil leaching from debris during temporary storage — Flushing of oil from debris with water	— Direct disposal — Burning — Degradation through land farming or composting for oil mixed with seaweed or natural sorbents
— Tar balls	— Separation from sand by sieving	— Direct disposal — Burning

Plastic bags   Plastic bags should be regarded as a means of transporting oily material rather than storage since they tend to deteriorate rapidly under the effect of sunlight, releasing their contents. It should also be borne in mind that if the contents are ultimately to be treated in some way prior to disposal, it will usually be necessary to empty the bags and dispose of them separately.

Separation techniques   Transport of recovered oily material can be expensive and, together with disposal, the costs can exceed those of collection. It is therefore beneficial to reduce the amount of material to be transported by separating oil from water and from sand during temporary storage. Water-in-oil emulsions can be broken to release the water; oil seeping from heaped beach material and debris can be collected in a ditch surrounding the storage area; and sieving techniques can be used to separate clean sand from tar balls.

Figure 7:
Plastic lined pit for temporary storage of collected oil/water mixtures.

Disposal Options

Recovery of Oils

Refineries   Under some circumstances it may be possible to recover the oil for eventual processing or blending with fuel oils. This should always be the first option to consider. Possible recipients for processing or blending are refineries, contractors who specialise in recycling waste oils, power stations, and cement and brick works. However, the quality of the oil **Quality criteria**   must be high since most plants can only operate with feedstocks meeting a narrow specification. For example, the oil should be pumpable, low in solids and have a salt content of less than 0.1% for processing through a refinery or less than 0.5% for blending into fuel oil. Small pieces of debris can be removed by passing the oil through a wire mesh screen. Assuming that the oil is suitable for recycling, it is likely that the potential **Inter-mediate storage**   refiners or other users will not have much spare storage or processing capacity and alternative intermediate storage may be required. Tanker deballasting stations and slop reception facilities may be appropriate in this regard but they may also have limited extra capacity.

Separation Techniques

Gravity separation

Oil collected from the water is likely to be the easiest to prepare for processing since it will usually only be necessary to separate any associated water. This separation can frequently be achieved by gravity either in collection devices such as vacuum trucks or in portable tanks, the water being run-off or pumped from the bottom of the tank.

Emulsions

Heat separation

The extraction of water from water-in-oil emulsions (mousse) is sometimes more difficult. Unstable emulsions can usually be broken by heating up to 80°C and allowing the oil and water to separate by gravity. In warm climates, the heat of the sun may be sufficient. More stable emulsions may require the use of chemicals known as 'emulsion breakers' or 'demulsifiers' which also tend to reduce the viscosity of most oils rendering them more pumpable. There is no single chemical suitable for all types of emulsion and it may be necessary to carry out trials on site to determine the most effective agent and optimum dose rate. However, typical dose rates are in the range of 0.1-0.5% of the bulk volume to be treated. Treatment is best carried out during transfer of the emulsion from the collection device to a tank or from one tank to another to ensure good mixing and therefore minimum dose rate. The emulsion breaker can be injected at the inlet side of a pump or into an in-line static mixer incorporated into a vacuum intake (Figure 8). After separation, the water phase will contain most of the emulsion breaker and up to 0.1% of oil and so care should be exercised over its disposal.

Emulsion breakers

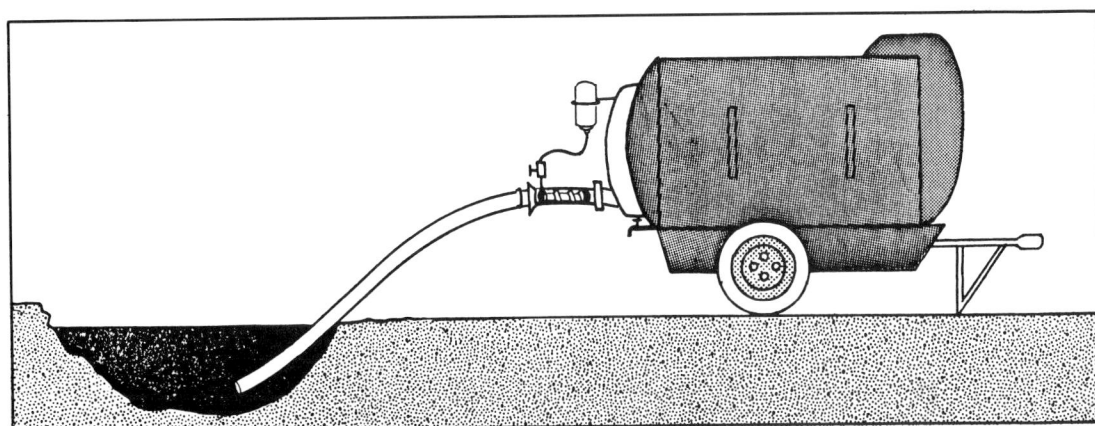

Figure 8:
Vacuum tank fitted with a dispenser for introduction of emulsion-breaker chemical into a static mixer incorporated in the suction hose.

Sand mixing

Recent experiments have suggested that emulsions can be partially broken by mixing thoroughly with sand in standard equipment such as concrete mixers. If an emulsion with 70% water is mixed with about 50% by volume of sand, the water content can be reduced by half and returned to the beach together with separated clean sand.

Recovery from beach material

On occasions it may be possible to recover oil from contaminated beach material. This usually involves washing the oiled beach material with water, sometimes in conjunction with a suitable solvent such as diesel to release the oil. Water washing using low pressure hoses can be used to loosen and lift off oil from debris contained in a temporary storage pit. The resulting oil/water mixture can then be pumped away and separated by gravity. Separation can also be achieved in a closed system using water or a solvent. Devices have been developed based on a range of readily available equipment from cement mixers for small-scale batch operations to mineral processing equipment for large-scale continuous treatment. Although these systems have proved successful in trials, they have not yet found widespread application at oil spill incidents. The cost of cleaning large amounts of oiled beach material on site could compare favourably with other methods that involve transporting the material some distance from the coast for subsequent disposal.

Direct Disposal

Landfill sites

A common disposal route adopted when recovery of oil is impractical is dumping in designated landfill sites. Materials intended for direct dumping should have a maximum oil content of about 20%. Sites should be located well away from fissured or porous strata to avoid the risk of contamination of ground water, particularly if this is abstracted for domestic or industrial use. Disused quarries and mines are often ideal.

Co-disposal with domestic refuse

The co-disposal of oil and domestic waste is often an acceptable method even though degradation of the oil is likely to be slow due to the lack of oxygen. However, oil appears to remain firmly absorbed by all types of domestic waste with little tendency to leach out. The oily waste should be deposited on top of at least 4 m of domestic refuse either in surface strips 0.1 m thick or in slit trenches 0.5 m deep to allow free drainage of water. The oily material should be covered by a minimum of 2 m of domestic waste to prevent the emergence of oil to the surface when subjected to compression from site vehicles. The total quantity of oil should not exceed 1.5% of the total volume of the site.

Beach burial

In the case of shorelines lightly contaminated with oily debris or tar balls, it may be possible to bury the collected material at the back of the beach well above high water mark provided there is no risk of damage to vegetation or that the oil could be uncovered through normal beach erosion. A covering of at least one metre should be sufficient.

Stabilisation

Binding agents

An inorganic substance such as quicklime (calcium oxide) can be used to bind oily sand, provided there are no large pieces of debris. An inert product is formed preventing the oil from leaching out. The stabilised material can be disposed of under less stringent conditions than unstabilised oily sand and can also be used for land reclamation and road construction where high load-bearing properties are not needed. If the material is to be used for construction, it is essential to compact it using road-building equipment.

Alternative materials

Quicklime

Although quicklime appears so far to be the best binding agent, other materials might also be applicable such as cement and pulverised fuel ash waste from coal fired power stations. There are also a number of commercial products that are based on the same raw materials but which have been treated to improve their efficacy. Practical experience at spills so far suggests that these are not as cost effective as the untreated raw materials. One advantage of quicklime over other materials is that the heat generated by its reaction with water in the waste reduces the viscosity of the oil which facilitates adsorption. Clearly the suitability of the technique is dependent upon a plentiful supply of stabilising material close to the spill location. Quicklime can usually be obtained from cement works.

Amounts required

Mixing or layering

The optimum amount of binding agent required is primarily dependent on the water content of the waste rather than the amount of oil and is best determined experimentally on site. Typically, for quicklime the amount required is between 5 and 20% by weight of the bulk material to be treated. Treatment can either be carried out using a mixing plant or a layering technique. The former, whilst offering better quality control and needing less land area, requires the use of expensive equipment including a continuous drum mixer. Smaller quantities can be treated in a batch process using standard concrete mixers. Provided there is sufficient land available close to the location of the spill, a layering technique is probably the most cost-effective. The waste is spread out to a depth of about 0.2-0.3 metres and the lime incorporated using a pulverising mixer.

Two-stage mixing

On occasions it may be preferable to carry out primary mixing in pits at the site of the spill to render the oiled material more suitable for transport. The final treatment can then be undertaken at a larger reception facility using specialised equipment. Inevitably, the above technique gives rise to a great deal of corrosive dust and, if possible, the treatment site should be selected so as to minimise its spread to adjacent property. It is also important that operating personnel wear protective clothing and face masks to protect skin, lungs and eyes.

Burning

Direct burning

When oil is burnt on open ground it tends to spread and be absorbed into the soil. In addition, a tarry residue may remain since it is rarely possible to achieve complete combustion. However, the direct burning of oily debris in open drums or other containers is a useful technique in remote areas where smoke will not be a nuisance.

Incineration

A number of portable incinerators have been developed which generate the high temperatures necessary for total combustion of oily waste. The rotary kiln and open hearth types are most appropriate for oils with a high solid content. As a general rule, incinerators used for domestic waste are not suitable since chlorides from seawater may cause corrosion. Industrial incinerators, whilst likely to tolerate salts, may not have sufficient capacity to deal with an additional burden created by a large quantity of oily waste. However, if long-term storage is available, this may be an acceptable route.

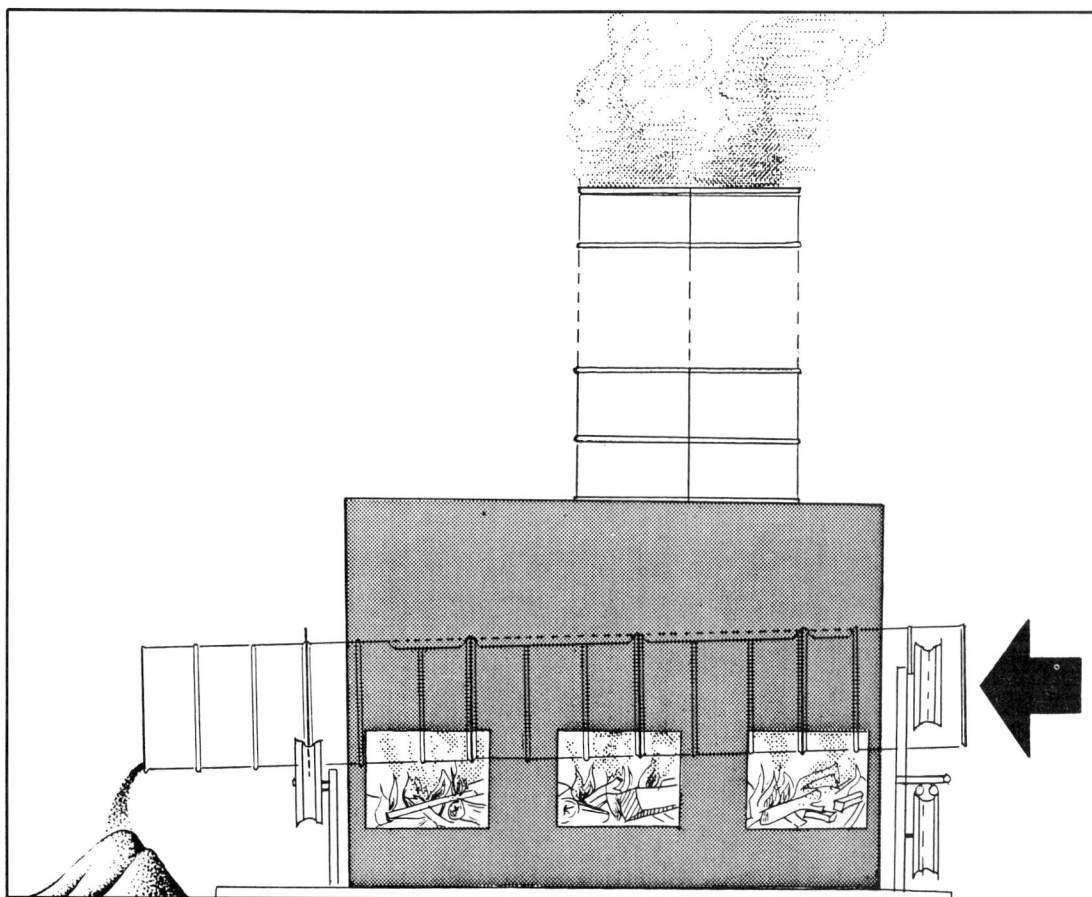

Figure 9:
Wood-fired kiln with inclined rocking cylinder constructed from 45 gallon steel drums.

Improvised kiln

One device developed for remote locations consists of a kiln which can be assembled on site from low cost materials such as 45 gallon steel drums. Oil-contaminated beach material is introduced manually at one end of the kiln at a rate of up to seven tonnes per hour and clean sand and pebbles are discharged at the other end (Figure 9). Combustion is self-sustaining if the feed material contains at least 25% oil and no more than about 50% water. The life-time of the unit may be quite short but should be capable of dealing with at least 100-600 tonnes of contaminated sand. A simpler portable burner suitable for the small-scale burning of tar balls and debris can be constructed from a single open 45 gallon drum. Air is supplied tangentially from a suitable compressor or fan blower to support combustion.

Portable burner

Enhanced Biodegradation

Limiting factors

Oil and oily wastes can sometimes be broken down using biological processes. Biodegradation of oil by micro-organisms can only take place at an oil-water interface, so that on land the oil must be mixed with a moist substrate. The rate of degradation depends upon temperature and availability of oxygen and appropriate nutrients containing nitrogen and phosphorous. Some oil components such as resins and asphaltenes are resistant to degradation and, even after prolonged periods, up to 20% of the original material may be left unaffected.

Commercial products

Nutrients

There are a number of products on the market which contain oil degrading bacteria and other micro-organisms. Some are intended for direct application to oil on shorelines together with nutrients to support the degradation process. Attempts to use these products in actual spills have met with very little success mainly due to the oil concentrations being too high, the lack of an oil-water interface and the difficulty in maintaining the required nutrient levels on a tidal shoreline. A more recent development which appears promising involves the addition of oil soluble nutrients to accelerate the process of natural degradation. These nutrients are more likely to remain at the oil-water interface rather than become dissolved in the sea.

Land farming

Application rates

A more effective approach is to distribute the oil and debris on land set aside for the purpose; a technique sometimes referred to as 'land farming'. In temperate climates, it may take as long as three years before the bulk of the oil is broken down, although degradation rates can often be increased by regular aeration of the soil and by the addition of fertilisers, such as urea and ammonium phosphate. The method is only likely to be applicable to relatively small spills because of the amount of land required. The contaminated material should not contain more than about 20% oil and ideally the land selected should be of low value, located well away from potable water supplies and should exhibit low permeability. The topsoil should first be loosened by means of a harrow and the area bunded to contain any oil run-off. The oily debris is then spread over the surface to a depth of no more than 0.2 metres, the maximum application rate being about 400 tonnes of oil per hectare of land. The oil should be left to weather until it is no longer sticky before being thoroughly mixed in with the soil using a plough or rotavator. Mixing should be repeated at intervals of 4-6 weeks for the first six months but less frequently thereafter.

Natural sorbents

If land farming techniques are employed, the use of natural sorbents such as straw and bark during clean-up is preferable to synthetic materials since they break down more rapidly. Large items of debris such as timber and boulders should be removed. Once most of the oil has degraded, the soil should be capable of supporting a wide variety of plants, including trees and grasses.

Composting

Another means of enhancing the degradation of small quantities of waste is to employ composting techniques, particularly if natural sorbents such as straw, peat or bark have been used. Provided the mixtures contain relatively low levels of oil, they can be stacked into heaps to facilitate composting. Because the heaps retain heat, the technique is particularly suitable in colder climates where degradation through landfarming is slow.

CONTINGENCY PLANNING

Manpower & equipment

Given the limited availability of resources, the protection and clean-up of all sites along a polluted length of coast may not be possible and priority areas must be identified in advance. Sources of both manpower and equipment must be established. Contractors who can provide lorries, vacuum trucks, front-end loaders, hot water units and other equipment need to be identified and the basis for hire charges established. Supplies of dispersant and sorbents should be located and adequate stocks held in areas where the risk of spills is high.

Communications

Although the men chosen to be supervisors will usually have local knowledge of the coastline, they will require proper training in the techniques and management of shoreline clean-up. Radio communications between the On-Scene Coordinator, individual beach supervisors and those responsible for storage and disposal should be provided.

Shoreline mapping

Decision guides presented as flow charts enable clean-up teams to think through problems before an event. Annotated maps of the coastline should show shoreline types, vehicle access points and beaches which would support heavy equipment. The same maps might also show coastal currents and prevailing winds, high priority areas, the location of environmentally sensitive resources and areas where dispersants cannot be used.

Disposal

Storage facilities

Plans for disposal should be specific to a given locality since the methods adopted will be largely dependent on the availability of raw materials and suitable disposal sites. Planners should establish the capabilities of local refineries and any oil recovery contractors, and obtain specifications regarding the quality of oil acceptable for processing. It is necessary to determine the availability and capacity of storage facilities and to select suitable sites for excavating pits. All disposal routes are likely to have limited capacities in relation to the rate at which the oil is collected.

Debris

Many disposal problems are directly related to debris. A survey of the coastline and the identification of debris collecting points will often indicate where spilled oil is likely to come ashore. Debris can sometimes be removed before the arrival of oil or, alternatively, a conscious effort might be made with the aid of booms to divert the oil away from the debris-laden area.

Exercises

Practical exercises of the overall contingency plan should be conducted periodically, not only to test the organisational aspects but also to ensure that the equipment and other resources identified in the plan are actually available and in working order.

POINTS TO REMEMBER

1. Shoreline type largely determines the most appropriate clean-up techniques.

2. In general, mobile oil should be picked up as soon as possible to prevent its movement elsewhere. The collection of stranded oil can often be left until all oil from a particular incident has come ashore.

3. Environmentally sensitive shorelines such as marshes, mud flats and mangroves are usually best left for natural cleaning processes to take place.

4. The use of mechanical equipment can clean beaches quickly, but substantial quantities of sand are also removed leading to disposal problems and potential erosion. Slower manual techniques are often better.

5. The effectiveness of clean-up should be closely monitored so that the techniques and level of effort remain in step with changing conditions, and to ensure that operations are terminated at the appropriate time.

6. In planning and executing a shoreline clean-up operation, a high level of organisation is required to make the best use of resources, especially where these are spread over a wide geographical area.

7. Sites for temporary storage should be identified which can provide a buffer between shoreline clean-up and final disposal.

8. The feasibility of recovering usable oil should be examined before other disposal techniques are adopted. Careful assessment of sites for direct dumping is essential to avoid ground water contamination.

9. Although a variety of techniques have been developed for disposing of oil and oily wastes, many have limited application and capacity. Selection of the most appropriate options should be made at the contingency planning stage. In the event of a major spill, all options will need to be considered.

FURTHER READING

API (1985) Oil spill response: Options for minimizing adverse ecological impacts. API, Washington, D.C., U.S.A. Publication No. 4398. 98 pp.

Breuel, A. (Ed.) (1981) Oil spill clean-up and protection techniques for shorelines and marshlands. Noyes Data Corporation, New Jersey, U.S.A. 404 pp.

CEDRE (1980) Effectiveness and costs of beach clean-up techniques and waste disposal. CEDRE, Brest, France. Report No. R.79.140.E. 25 pp.

CONCAWE (1980) Sludge farming: a technique for the disposal of oily refinery wastes. CONCAWE, The Hague, Netherlands. Report No. 3/80. 94 pp.

CONCAWE (1980) Disposal techniques for spilt oil. CONCAWE, The Hague, Netherlands. Report No. 9/80, 52 pp.

CONCAWE (1981) A field guide to coastal oil spill control and clean-up techniques. CONCAWE, The Hague, Netherlands. Report No. 9/81. 112 pp.

CONCAWE (1984) Capability of oil industry installations for the disposal of spilt oil. CONCAWE, The Hague, Netherlands. Report No. 8/84. 38 pp.

Exxon (1984) Oil spill response field manual. Exxon Production Research Co., Houston, U.S.A. 137 pp.

IMO (1980) Manual on oil pollution — Section IV Practical information on means of dealing with oil spillages. IMO, London. 143 pp. (under revision).

ITOPF (1983) Shoreline clean-up. Technical Information Paper No. 7. ITOPF, London. 8 pp.

ITOPF (1984) Disposal of oil and debris. Technical Information Paper No. 8. ITOPF, London. 7 pp.

WSL (1982) Oil spill clean-up of the coastline: a technical manual. Warren Spring Laboratory, Stevenage, U.K. 72 pp.

V PLANNING AND OPERATIONS

Careful planning is essential if an emergency such as an oil spill is to be dealt with success-fully. Many people may be affected by an oil spill and many organisations will have duties to perform apart from the task of physical clean-up.

This section provides guidance for the preparation of contingency plans. It is suggested that the first part of a plan should outline the overall strategy for oil spill response whilst the second half should give the operational procedures to be followed when a spill occurs. In order to illustrate how a contingency plan works in practice, the events of a hypothetical oil spill are summarised within the description of an Operational Plan.

CONTENTS

SCOPE AND CONTENT OF CONTINGENCY PLANS

Tiered response

Most oil spills are small and can be dealt with locally. Should the incident prove beyond the local capability or affect a larger area, an enhanced but compatible response will be required. The foundation of this tiered response is the local plan for a specific facility such as a port or oil terminal or for a length of coastline at risk from spills. These local plans may form part of a larger district or national plan. National plans may in turn be integrated into regional response arrangements covering two or more neighbouring countries.

Common format

In general, contingency plans should follow a similar layout irrespective of whether they are local, national or regional in scope though their length and content will vary with the size of the area covered and degree of risk. Similarity in layout will enable the plans to be easily understood, will assist compatibility and ensure a smooth transition from one level to the next.

Two parts

Up-dating

Contingency plans can be divided into two main parts under the headings STRATEGY and OPERATIONAL PLAN. The strategy segment of the plan should define the policy, responsibilities and rationale for the operational plan which is essentially an action checklist with pointers to information sources (Figure 1). A plan should be reasonably complete in itself and should not entail reference to a number of other publications, which causes delay. A loose-leaf format facilitates regular updating and there should be provision for listing and dating amendments.

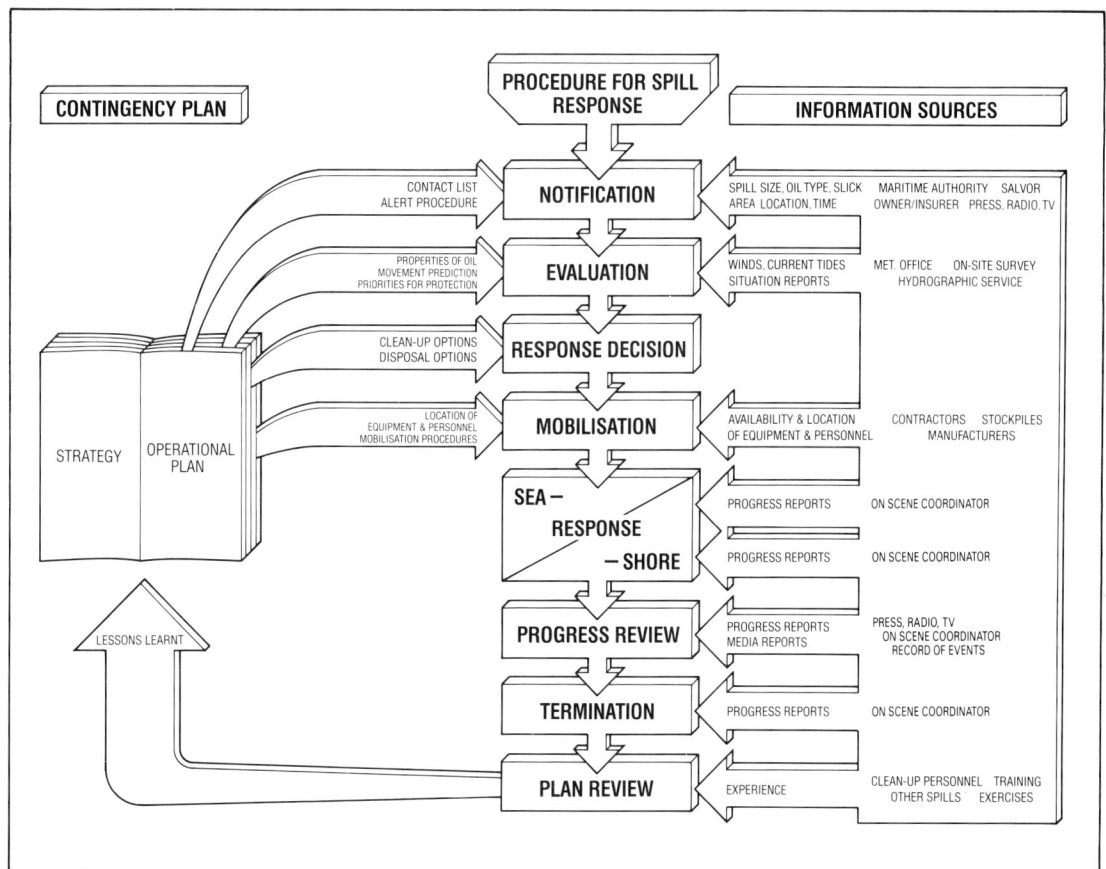

Figure 1   *The role of contingency planning in oil spill response*

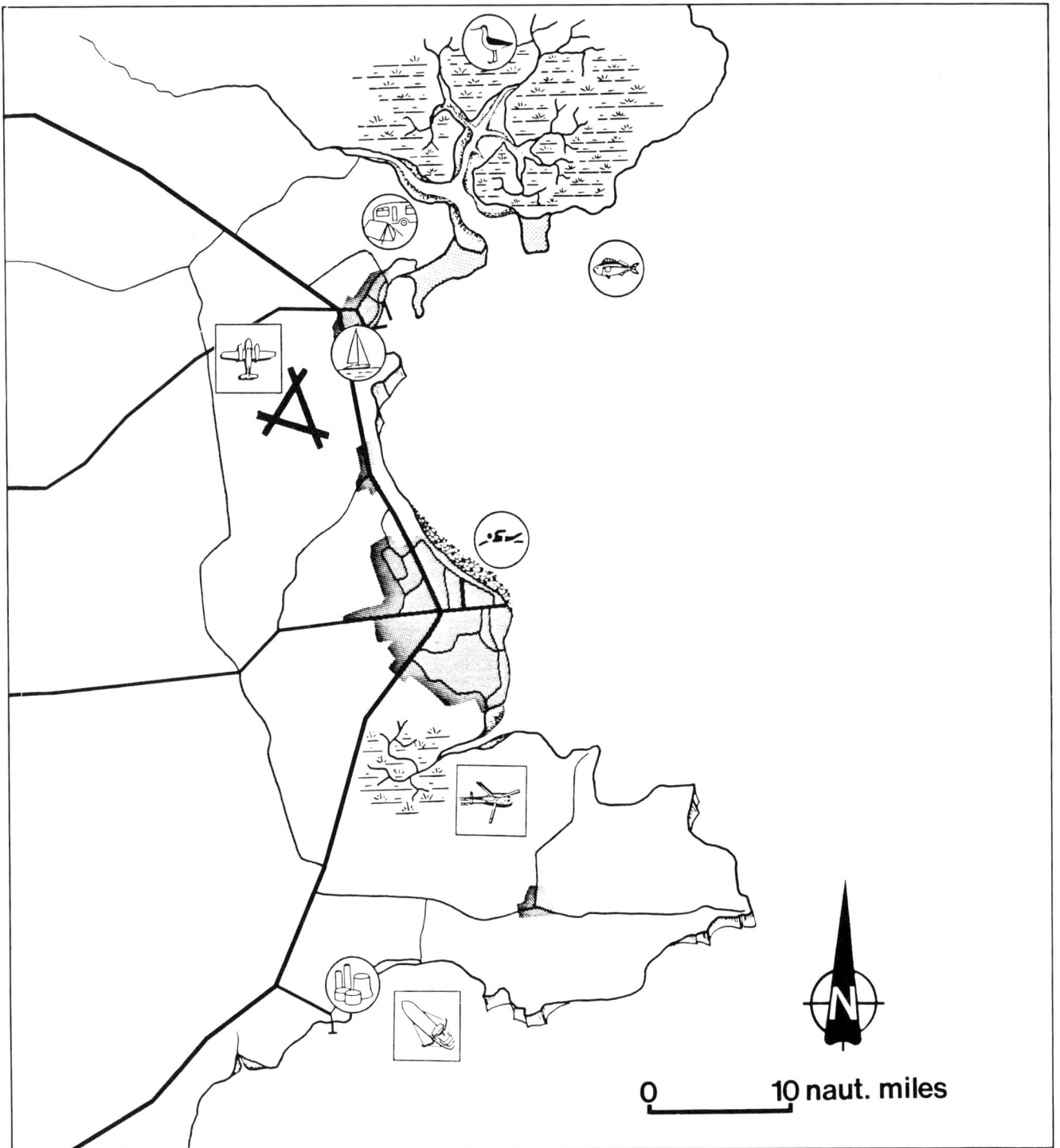

	MUD		CAMP SITE		WATER INTAKE
	SAND		YACHT MARINA		HELICOPTER BASE
	WETLANDS		BATHING BEACH		AIRFIELD
	CLIFFS		BIRDS		EQUIPMENT DEPOT
	PEBBLES		INSHORE FISHERIES		

0   10 naut. miles

STRATEGY

Introduction

Responsi-bility

Responsibility for and the scope of the contingency plan should be defined in the introduction. This should identify the authority or lead agency responsible for the formulation and implementation of the plan and explain the statutory requirements, if any, upon which this responsibility is based. The geographical area covered by the plan should be defined and reference made to any other related plans.

Assessment of Spill Risk

Past spill records

Number of port calls

The expected frequency and size of spills, and the types of oil likely to be involved should be addressed in this section. Historical spill records for the area covered by the plan may be helpful but, because spills occur infrequently, there is often insufficient data to make a fully quantitative assessment. If such records are unavailable, it is usually possible to make comparisons with other locations where they do exist, taking into account any differences in circumstances. For oil ports, the number of calls made by tankers is relevant in assessing risks since most spills from tankers are small and occur in these locations as a result of routine operations such as loading, discharging and bunkering. A range of possible spill scenarios can be developed from an analysis of oil-related activities in the area and types of oil moved or handled.

Movement and Persistence of Oil

Oil charac-teristics

The probable fate of oil slicks should be analysed in relation to the types of oil likely to be spilled and the prevailing seasonal weather conditions. A note of the physical properties of any such oils, in particular specific gravity, viscosity, pour point at usual sea temperatures and distillation characteristics should be annexed to the operational plan.

Winds and currents

Computer models

The movement of oil can be predicted from a knowledge of the strength and directions of tides, currents and winds. Where the potential source of a spill can be identified, the most likely movement of the oil can be calculated and any seasonal variations in the direction of prevailing winds taken into account. Reference to any computer models designed to predict movement of oil would also be made here. Information on tidal streams or residual currents would form an annex to the operational plan.

Figure 2   *(opposite) Resource maps*

The scale of the maps and the number required will depend upon the size of the area covered by the contingency plan and the complexity of the features to be illustrated. Those accompanying district and national plans will usually only give a broad indication of the main features of the coastal region, the resources at risk and potential sources of oil spills. Those accompanying local contingency plans for a restricted length of coastline will also provide more detailed information, such as the probable movement of surface slicks, agreed response strategy, shore access points and temporary storage and disposal sites. For clarity it may be appropriate to divide information between two or more maps. Reference may also be given to additional sketches or photographs which illustrate elements of the response arrangements in more detail. Maps should not be regarded as a substitute for written text but as a means of illustrating key points.

Resources at Risk from Oil Spills

Sensitive resources

Maps

Amenity areas, ecologically sensitive areas, industrial sea water intakes, fisheries, mariculture, seabirds, marine mammals and other resources likely to be threatened should be identified. A summary of the important features should be included in this section while detailed information on the location of each resource should be annexed to the operational plan. This is often done with the aid of maps to indicate the location of sensitive resources and the priorities for protection (Figure 2). Such maps should not be limited to biological resources; sensitive industrial and recreational resources should also be shown.

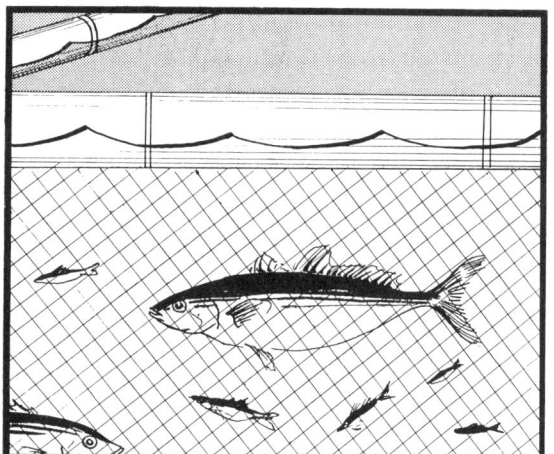

Limited capability

Realistic aims

Priorities for protection must be determined since in a major spill it is unlikely that all the resourses at risk can be successfully defended. This is probably the most important facet of the policy adopted for spill response and only governmental authorities are in a position to make the necessary decisions, since the economic and environmental values to the community will have to be assessed. However, it is essential to take into account not only how desirable protection of a particular resource would be, but also to what extent its defence is practicable. Provision should also be made for response priorities to be altered if resources are impacted by a spill before the plan can be implemented.

Seasonal variation

Seasonal variation can greatly alter priorities. For example, the high priority given to an amenity beach in summer may not apply in winter. Similarly, certain biologically sensitive areas may be assigned high priority during breeding seasons or when migratory species are known to be present. The maps denoting sensitive areas attached to the operational plan should be clearly annotated with such information.

Selection of Techniques

Limitations
of
techniques

The clean-up strategy should be determined in relation to the assessment of the risk of spills and to the defence of agreed priorities for protection. The limitations of spill control techniques must be appreciated and the most suitable equipment selected for the anticipated range of weather conditions and oil types. As an example, boom deployment sites should be designated only where containment and recovery of oil or its deflection to less sensitive areas is actually feasible. The different shoreline types falling within the area covered by the plan should be identified and the most appropriate clean-up strategy

Shoreline
types

for each considered. Factors to be taken into account include its amenity value, whether beaches are easily accessible for heavy equipment and the ability of the beach to support such vehicles. Maps attached to the operational plan can be used to show the areas where each technique such as dispersant spraying should be used and where any restrictions might apply.

Maps

Location of
equipment

The location of equipment adjacent to high risk areas ensures a rapid and effective response. However, a balance must be found between stockpiling at central points with the inherent transport costs and delays, and the more expensive option of equipment packages at every potentially vulnerable site. Procedures for mobilisation must be set out in the operational plan while an inventory of available equipment should be annexed. Descriptions are most easily presented as a table where details such as type, dimensions, capacity, transport requirements and a contact point for its release are listed against location. Cross-reference systems permit the locations of all equipment of a particular type to be identified, and lend themselves to computer techniques. In some cases equipment and services will be owned or provided by contractors, industry or other parties, which makes it desirable to define in an annex to the plan, the contractual terms acceptable in principle to the parties concerned.

Inventories

Contracts

Manpower requirements

The manpower required to deploy the equipment and undertake clean-up will need to be estimated. The extent to which the requirement can be met from the organisation implementing the plan will depend upon availability, the techniques involved and the amount of specialised equipment to be deployed. In the case of large spills, additional manpower may be required, particularly for labour-intensive operations like shore clean-up. Sources of back-up labour such as contractors and government departments should be listed in an annex to the operational plan.

Temporary storage sites

Final disposal options

Temporary storage sites and disposal routes for oily wastes must be agreed in advance. Locations close to areas of risk suitable for temporary storage of oil and oily waste have to be identified. The disposal options should be discussed and a decision made, taking into account the environmental considerations of each method and the probable costs of transport and disposal. Details of the disposal methods selected should be annexed to the operational plan. Temporary storage sites are best shown on the maps delineating shoreline clean-up techniques.

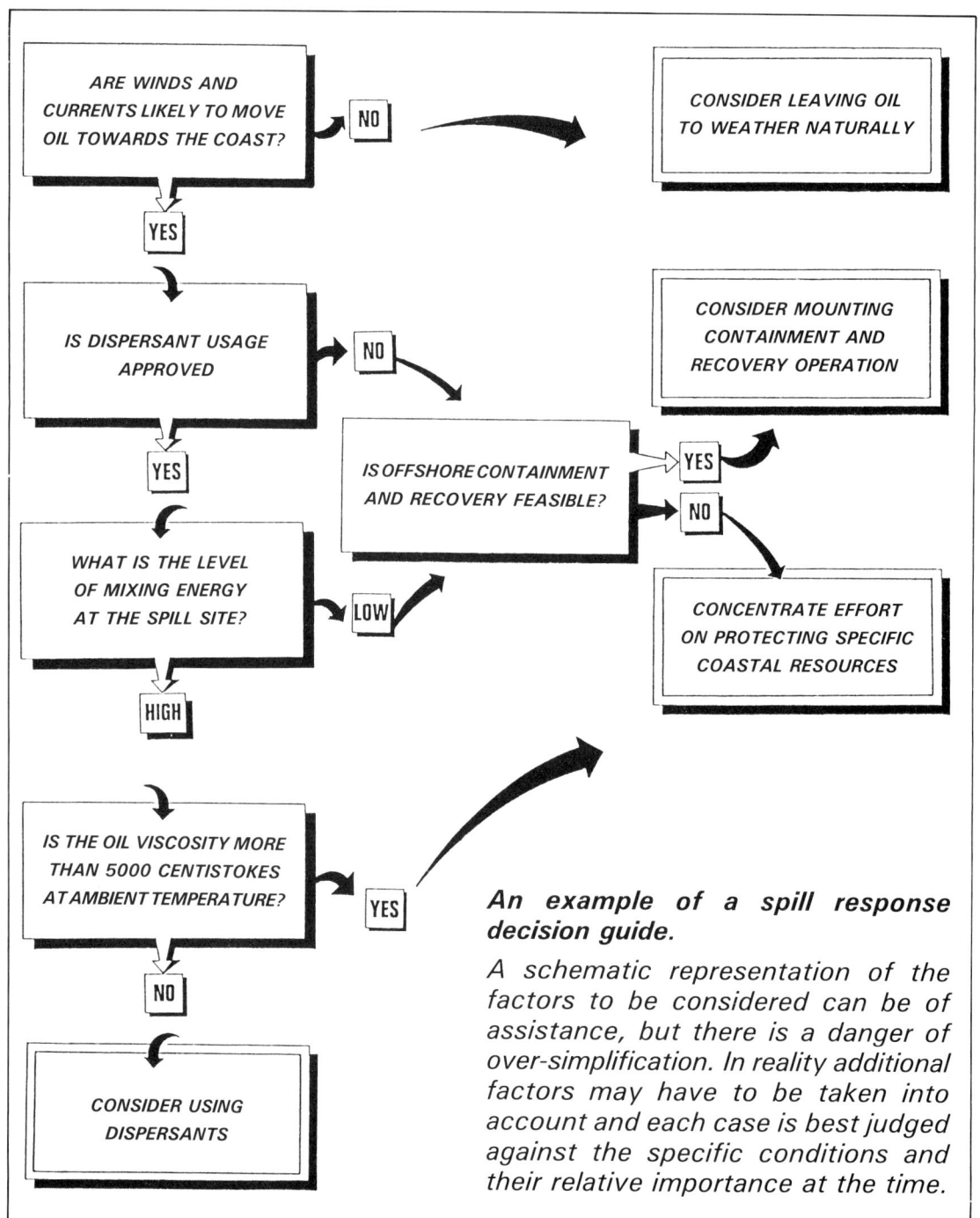

An example of a spill response decision guide.

A schematic representation of the factors to be considered can be of assistance, but there is a danger of over-simplification. In reality additional factors may have to be taken into account and each case is best judged against the specific conditions and their relative importance at the time.

Outline of the Response Organisation

Division of responsibility

Responsibility for oil spill control at sea usually rests with a government agency involved in maritime affairs such as a coast guard or navy. Only rarely do the responsibilities of such agencies extend to the shoreline. Generally the task of cleaning inshore waters and shorelines falls to port authorities or local government. This division of responsibility is apt to cause confusion.

Co-ordination

Central co-ordination under a single organisation which has overall responsibility for handling the operation is normally the best solution. Where this is not possible and more than one organisation is involved in responding to a spill, procedures for co-ordination between the various groups must be laid down. In a major spill, the on-scene co-ordinator or commander will delegate control of operations, whilst in a smaller incident co-ordination and control functions may be combined. The size of the organisation depends on the area covered by the plan, the severity of the threat and the sensitivity of the resources threatened. Clean-up of shorelines involves more interests and typically demands greater co-ordination than response at sea. In every case, responsibilities should be clearly defined and, as far as possible, the number of authorities involved kept to a minimum.

Communications

Radio frequencies

A communications centre is essential, with telephone, telex and radio communications to ensure that the necessary information is passed to the appropriate people. The centre serves as the ideal focal point throughout the response operation since all information on clean-up and logistic support will be channelled through it. In a major spill, operations at sea, on shore and in the air will be taking place at the same time. In addition to a common radio frequency, it may be necessary to allocate separate frequencies for each operation. Where clean-up operations are conducted over extended distances, portable communications centres should be located close to the scene of each operation. Repeater stations may be required so that communications can be maintained.

Logistic support

Customs and immigration

Logistic support is an essential element in a contingency plan to ensure that the clean-up operation runs smoothly. Arrangements for providing food, clothing, shelter and medical support to shore clean-up crews must be considered in advance. The availability of back-up resources, such as additional equipment, materials, and transport, should also be examined together with the names and addresses of potential suppliers, both within the country and from neighbouring countries. In the latter case attention must be given at the planning stage to immigration and customs clearance procedures. Delays may result from normal immigration and customs formalities and the plan should provide for urgent clearance in an emergency when personnel and equipment need to be brought into a country.

Record keeping

Documentation of actions is important and accurate records should be kept regarding the use of manpower, equipment and materials, and the related expenditure. For the sake of consistency, it is worth preparing examples of record forms and annexing these to the operational plan. Good documentation will assist in formulating claims when the operation is completed.

Liaison

Public relations

Liaison arrangements with other interested parties such as government authorities and organisations not immediately involved in the response operation, but with interests in certain facets of the spill, should be included in the plan. Examples include operators of industrial plants which abstract sea water, environmental protection groups and other government departments. It is often useful to set up a committee to keep all such parties up to date with contingency planning and for consultation during a major spill. Provision should be made for keeping the news media informed during a spill without interfering with the conduct of the operation. This is likely to require additional telephone lines, separate accommodation and well-informed briefing officers.

ADVICE	OPERATIONS	SUPPORT

```
ADVICE                OPERATIONS                    SUPPORT

Salvage          ┌──────────────────┐        Communications
Firefighting     │ On Scene Coordinator │      Administration
Technical        └──────────────────┘        Public Relations
Scientific                                    Logistics :-
Legal         ┌──────────────┬──────────────┐      Vehicles
Insurance     │ Shore Supervisor/s │ Sea Supervisor/s │   Vessels
              └──────────────┴──────────────┘      Aircraft
                                                   Equipment
          ┌──────────────┐  ┌────────┬────────┐    Materials
          │ Beach Team Leaders │ │ Vessel │ Aircraft │  Manpower
          └──────────────┘  │ Operators │ Operators │ Provisions
                            └────────┴────────┘     Accommodation
```

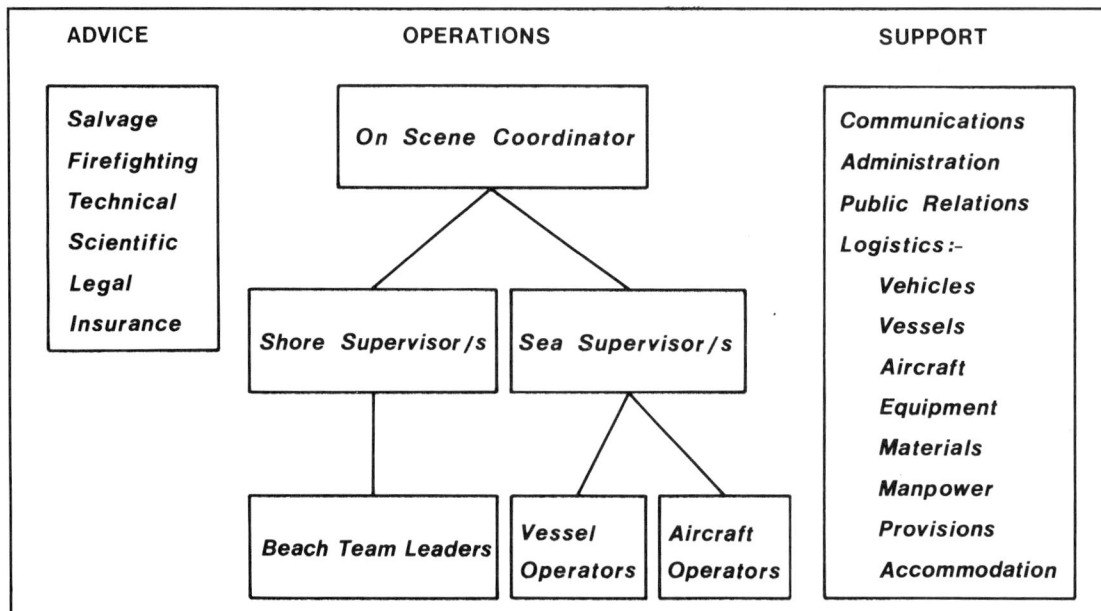

Organisation chart for oil spill response. *In the case of minor spills, roles may be combined.*

Training and Review Procedures

Training

Testing of equipment

Procedures for training and exercises, and for updating the plan should be defined. Training programmes should be developed at all levels including personnel in charge of boats and shore clean-up parties. Regular exercises will ensure that contingency arrangements function properly and that all those likely to be involved in a spill become fully familiar with their particular responsibilities. From time to time equipment listed in inventories should be mobilised and deployed to test its actual availability and performance.

Up-dating plan

An oil spill provides the best oportunity for improving a contingency plan. Events should be reviewed soon after the clean-up operation has been terminated and the plan revised on the basis of lessons learnt, when memories are still fresh.

OPERATIONAL PLAN

The operational plan should describe the recommended procedures for responding to an oil spill with essential information included as annexes. Many events during an oil spill response operation will occur concurrently but the format of the operational plan should follow the roughly chronological order indicated in the following sequence.

Notification

Four o'clock in the morning. On-scene Commander notified by Coast Guard that tanker aground 10 miles off coast and two cargo tanks damaged containing total of 10,000 tonnes of crude oil. On-scene Commander telephones other Response Committee members and arranges meeting in Incident Room being established at local Coast Guard station. Instructs Coast Guard to obtain update from tanker master and to pass initial reports to primary contacts using agreed alert procedure laid down in contingency plan.

The first information regarding an oil spill may come from any one of a number of sources including the general public. The police and other emergency services must hold telephone numbers, telex numbers and radio frequencies allowing them to contact the agency designated to receive such information on a 24-hour basis, e.g. coast guard, military, marine or fire services. On receipt of this information, the designated agency should transmit an initial report as soon as possible to interested parties according to an agreed alert procedure. The format of such a report should be included in the plan and contain the following:

— *Date and time of observation (specifying local time or GMT).*
— *Position (e.g. latitude and longtitude or nearest landmark).*
— *Source and cause of pollution, (e.g. name and type of vessel; collision or grounding).*
— *Estimate of amount of oil spilled and likelihood of further spillage.*
— *Description of oil slicks including direction, length, breadth and appearance.*
— *Type of oil spilled and its characteristics.*
— *Weather and sea conditions.*
— *Action, both taken and intended, to combat pollution and prevent further spillage.*
— *Name and occupation of initial observer and any intermediate reporter and how they can be re-contacted.*

It should be made clear that the initial report should not be delayed as long as the first three headings can be satisfied, the remainder being transmitted as soon as available.

Evaluation

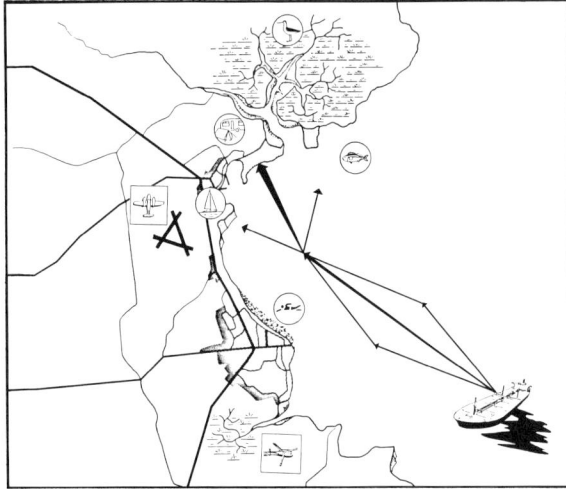

Response committee meet and assess what is known. Master reports 2000 tonnes of Light Arabian crude oil have been spilt so far. Weather reports and current data collated and coastlines at risk identified from contingency plan. Knowledge of characteristics of oil allows prediction of its probable fate, including its drift and rate of natural dissipation. Observations of appearance and drift of oil on sea relayed from surveillance aircraft confirms that oil poses serious threat to amenity beaches, harbour, estuary and marshland with its migratory bird populations. Arrangements made to advise interested parties and on-scene co-ordinator calls meeting to discuss implications.

The plan should provide for the on-scene co-ordinator (or duty staff member) to evaluate the situation and to assess the threat posed by the oil to the resources at risk. Action should be taken to:

— *Identify the type of oil in terms of specific gravity, viscosity, pour point, wax content and distillation characteristics.*

— *Determine the expected track of the oil slick at regular intervals from data on currents, tides and winds.*

— *Consider arranging on-site surveillance using aircraft to verify predictions and obtain further details.*

— *Identify threatened resources.*

— *Inform the parties who might be affected by the spill.*

Sources of the information required to make the necessary evaluation include the plan, the master, salvor, insurer and port authority.

Response Decisions

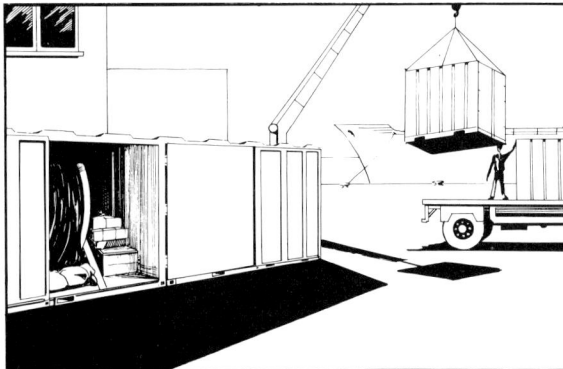

In view of threat, decision made to mount active response. Instruction passed to modify fixed-wing aircraft for spraying dispersant, to prepare spray buckets for use by helicopters and to place offshore oil recovery vessels on-standby. Booms also to be made ready for transport to priority protection sites and beach clean-up equipment prepared for mobilisation. Additional equipment from other regions and neighbouring countries not requested. Responsibilities delegated to sea and shore supervisors.

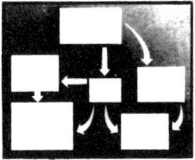

The plan should provide for the various response options to be considered.

— *If no key resources are threatened, no response may be necessary beyond monitoring the movement and behaviour of the slick.*

— *If key resources are threatened, decide whether their protection is best achieved by combating the oil at a distance or by the use of booms or other measures to defend specific sites.*

— *If no protection is feasible or if resources have already been affected, decide on the priorities for clean-up.*

— *Select the necessary equipment and manpower required and determine availability and location.*

Arrangements should be included in the plan for placing manpower and equipment on stand-by: equipment may be loaded onto vehicles ready for despatch and paperwork completed before the actual mobilisation order is given.

Clean-up Operations

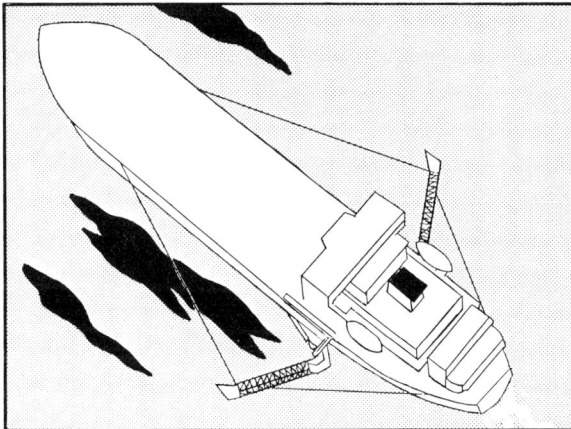

At 9 o'clock planes instructed to commence spraying operations, and vessels to attempt recovery of oil not being sprayed. Boom deployment on coastline commenced since evident that operation at sea will not be totally successful. At harbour, boom deployed successfully but previously untested estuary and shingle beach plans run into problems due to weather and tidal conditions and inadequate equipment. Helicopters with spray buckets mobilised as oil gets nearer to shore. Repeated aerial surveillance carried out to monitor oil movement and advise progress of response at sea.

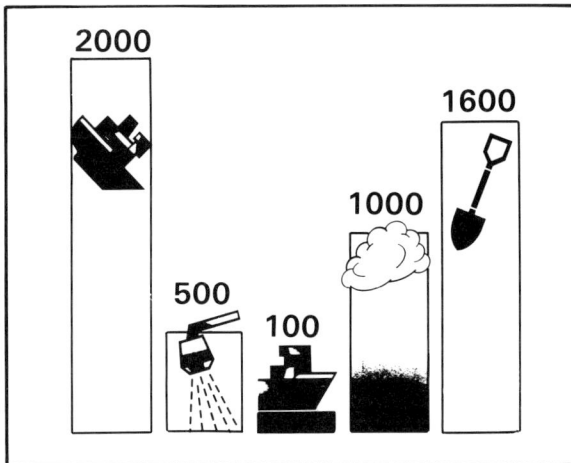

During afternoon spraying operations terminated as oil too viscous. Recovery vessels continue sweeping windrows guided by control aircraft until dusk when return to port to discharge oil. By end of first day estimated that of 2000 tonnes of oil originally spilled, 1000 tonnes has evaporated and dispersed naturally, 500 tonnes has been chemically dispersed and 100 tonnes recovered by vessels. Remaining 400 tonnes has taken up four times its weight of water to form 1600 tonnes of "mousse"; almost as much as the quantity of oil originally spilled, despite relatively successful clean-up operations at sea.

On day two most of the remaining oil comes ashore and effort is concentrated on shore clean-up. Oil contained within boom at harbour removed by skimmers. Shingle amenity beach only affected by isolated patches of oil but sand bar and marshland affected quite severely. Decision made to remove oil from sandy shoreline using manual methods with vehicles in support. Against all previously agreed policy, marsh clean-up carried out because of threat to migratory birds. Methodical clean-up of sand beaches by small teams proves relatively straightforward but problems of access, damage by vehicles and difficulties of removing oiled vegetation at marsh site demonstrate why clean-up of such areas was not recommended in contingency plan.

Procedures should be laid down for:

— *Mobilising the necessary equipment and related manpower.*

— *Deploying equipment at sea and on shore in accordance with the response decision and placing booms at pre-designated sites to protect key resources, referring to details of mooring points and configurations.*

— *Organising sufficient logistic support so that there are no bottlenecks, (e.g. between oil collection, temporary storage and final disposal) and arranging for supply of dispersants, fuel, food, clothing and other consumables.*

— *Using aircraft to control clean-up operations at sea and to maintain overall surveillance of spill, both at sea and on shore.*

— *Selecting the most suitable disposal route depending on the nature of any collected oil.*

— *Reviewing the progress of the clean-up operation using inputs from aerial surveillance and personnel on the site to reassess the response decisions.*

— *Maintaining accurate records, on a daily basis for each clean-up location, of all the actions taken, manpower and equipment deployed, amounts of materials used.*

Communications

The response operations are aided by good communications. The incident room is equipped with radio to talk directly to vessels and aircraft engaged in the operation and to relay information to the mobile command post. It is also served by adequate telephone lines. The surveillance aircraft and helicopters are able to communicate directly with the spray aircraft and recovery vessels. On the shore, the mobile command post and radio communications ensure a good flow of information along the chain of command, between the On-scene Commander, the shore supervisors and the foremen of the small work teams. Good communications with the public ensured through preparation of informative briefs for media.

This section of the plan should provide for:

— *Locating communications/command post as close to the scene of the spill as possible, ensuring that the entire area affected by the spill is within easy reach by radio or telephone.*

— *Ensuring that supervisory staff have the necessary radio equipment and are familiar with communications procedures, telephone, telex and telefax numbers, radio frequencies and call signs.*

It is essential that sufficient communications equipment should be available to allow the rapid transfer of information and instructions between aircraft, vessels, vehicles, shore clean-up parties, and the central communications and command post.

Termination of Clean-up

After a few days clean-up of the marsh terminated since operation doing more damage than oil. Although most oil removed from sand beaches, approaching tourist season dictates further clean-up of small tar balls that will melt as temperature increases. On exposed shores not visited by public any remaining oil will be left to degrade naturally. Attention directed to restoring areas used for temporary storage of recovered oil and debris and to arrange final disposal for different wastes. Equipment returned to store for cleaning and repair and replenishment of stocks of dispersant and other consumables ordered. Detailed report of operation and collation of daily work records commenced in order to support claims for compensation.

While it is important to terminate an operation when it becomes ineffective or when the desired level of clean-up has been achieved, it is difficult to give precise guidance on this in a contingency plan. Provision should, however, be made for:

— *Liaison with all interested parties regarding the conduct of the operation and the level of clean-up appropriate to each location.*

— *Standing down equipment and returning it to stores for cleaning and maintenance. Re-ordering consumed materials and repairing or replacing damaged equipment.*

— *Restoring temporary storage sites and tidying up other work areas.*

— *Preparing a detailed report on the operation which can be used to support any claims for clean-up expenses and to review the contingency plan.*

V.15

Formulation of Claims

Clean-up effort expended, equipment used and materials consumed:-

at sea: 3 recovery vessels, 3 fixed wing aircraft, 2 helicopters, 2 bucket spray units, 25 tonnes dispersant;

at shingle beach: 40 mandays, 800 metres boom, 2 steam cleaning units;

at sandy beach: 800 mandays, 8 front-end loaders, 4 dumper trucks, sand-cleaning machine, 8 lorries, plastic sheeting, 25 tonnes quicklime;

on marsh: 240 mandays, 3 tractors, 1 dumper truck, 650 metres boom, 250 metres metal roadway;

in harbour: 80 mandays, 300 metres boom, 1 disc skimmer, 1 rope mop skimmer, 2 vacuum trucks, 2 steam-cleaning units, 5 drums dispersant, 2 bales sorbent boom.

In total, 4000 tonnes of oily waste collected in temporary storage for final disposal.

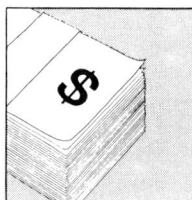

Compensation for clean-up costs and oil pollution damage from shipping accidents is normally available through insurance arrangements provided by shipowners' mutual Protection and Indemnity Associations (P & I Clubs). In the case of tanker incidents additional funds contributed by oil cargo owners may become available subject to certain conditions.

Claims may be made for costs incurred in prevention and control of oil pollution, for direct economic loss and costs for replacement and repair of property. Ideally claims should contain the following information in an appropriate language:

— *Name and address of the claimant.*

— *Date and location of the incident.*

— *Identity of the source involved.*

— *Type, quantity and geographical impact of spilt oil.*

— *Evidence in the form of chemical analysis, or movement of floating oil, to establish the original source of pollution and link it with the claim.*

— *Summary describing: the sequence of events of the incident; the actions taken in response to it; and the extent of damage to property or activity.*

— *A break-down of costs related to specific clean-up activities at each main work site, e.g. labour, equipment, material, and transport costs on a daily or weekly basis.*

— *Costs incurred or anticipated in replacing or repairing damaged items of declared age;*

— *Comparitive figures of previous earnings and lost profits to prove economic loss.*

— *Maps and photographs illustrating the above points.*

Claims handling can be slow and complicated involving lengthy correspondence, but much can be done by claimants to accelerate the process by recording events and expenditure and submitting clear and detailed claims that may be understood by claims handlers who may have no first-hand knowledge of the incident. Sources of further information on this subject are given at the end of this section.

TEN QUESTIONS FOR ASSESSING CONTINGENCY PLANS

1. Has there been a realistic assessment of the nature and size of the possible threat, and of the resources most at risk, bearing in mind the probable movement of any oil spill?

2. Have priorities for protection been agreed, taking into account the viability of the various protection and clean-up options?

3. Has the strategy for protecting and cleaning the various areas been agreed and clearly explained?

4. Has the necessary organisation been outlined and the responsibilities of all those involved been clearly stated — will all who have a task to perform be aware of what is expected of them?

5. Are the levels of equipment, materials and manpower sufficient to deal with the anticipated size of spill. If not, have back-up resources been identified and, where necessary, have mechanisms for obtaining their release and entry to the country been established?

6. Have temporary storage sites and final disposal routes for collected oil and debris been identified?

7. Are the alerting and initial evaluation procedures fully explained as well as arrangements for continual review of the progress and effectiveness of the clean-up operation?

8. Have the arrangements for ensuring effective communication between shore, sea and air been described?

9. Have all aspects of the plan been tested and nothing significant found lacking?

10. Is the plan compatible with plans for adjacent areas and other activities?

FURTHER INFORMATION

Abecassis, D.W. and Jarashow, R.L. (1985) Oil pollution from ships. 2nd Ed. Stevens & Sons, London. ISBN 0 420 47000 X. 619 pp.

Gold, E. (1985) Handbook on marine pollution. Assuranceforeningen Gard, Arendal, Norway. 247 pp.

IMO (1978) Manual on oil pollution, Section II (Contingency Planning). IMO, London. 52 pp.

IOPC Fund (1982) Claims manual. IOPC Fund London. 6pp.

ITOPF (1985) Contingency planning for oil spills. Technical Information Paper No. 9, ITOPF, London. 8pp.

ITOPF (1986) Action: oil spill. Technical Information Paper No. 12, ITOPF, London. 8 pp.

Useful addresses:

International Maritime Organization
4 Albert Embankment
London SE1 7SR

Telephone: (01) 735 7611
Telex: 23588
Fax: (01) 587 3210

International Oil Pollution Compensation Fund
4 Albert Embankment
London SE1 7SR

Telephone: (01) 582 2606
Telex: 23588
Fax: (01) 587 3210

Cristal Limited
Staple Hall
Stonehouse Court
87-90 Houndsditch
London EC3A 7AB

Telephone: (01) 621 1322
Telex: 888043
Fax: (01) 626 5913

International Tanker Owners Pollution Federation Limited
Staple Hall
Stonehouse Court
87-90 Houndsditch
London EC3A 7AX

Telephone: (01) 621 1255
Telex: 887514
Fax: (01) 626 5913

ACKNOWLEDGEMENTS

ACKNOWLEDGEMENTS

The International Tanker Owners Pollution Federation acknowledges the assistance provided by numerous individuals and organisations in the preparation of the series of Technical Information Papers upon which this book is based. We thank particularly:

J.M. Baker, Field Studies Council, UK
Bermuda Biological Station
Biggs Wall Fabricators, UK
E. Blomberg, Sweden
C. Bocard, Institut Francais du Petrole, France
British Petroleum Group Environmental Services, UK
J.A. Butt, Shell International Marine, UK
G. Canevari, Exxon Research & Engineering, USA
CEDRE, France
Chevron Shipping Company, USA
J. Churchill, Harvest Air, UK
R.B. Clark, University of Newcastle, UK
C.R. Corbett, United States Coast Guard
CRISTAL, UK
Department of Agriculture and Fisheries for Scotland, UK
A.C. Dykes, Thomas Miller P & I, UK
B. Emery, Conair, Canada
Esso Petroleum, UK
J.A. Galt, National Oceanic and Atmospheric Administration, USA
C. Getter, USA
E. Gilfillan, Bowdoin College, USA
L. Giulini, USA
M.S. Greenham, Canadian Coast Guard
E.R. Gundlach, USA
T.M. Hayes, International Maritime Organization
S. Hope, IPIECA, UK
I.W. Hughes, Department of Agriculture and Fisheries, Bermuda
IVL — Swedish Environmental Research Institute
International Oil Pollution Compensation Fund
T.G. Jacques, Ministere de la Sante Publique et da la Famille, Belgium
B. Koons, Exxon Production Research, USA
P. Korver, Netherlands

H.W. Lichte, USA
J. Lindstedt-Siva, Atlantic Richfield, USA
D. Mackay, University of Toronto, Canada
C.D. McAuliffe, Chevron Oil Field Research, USA
Marine Pollution Control Unit, UK
Micronair (Aerial), UK
Ministry of Agriculture, Fisheries & Food, UK
M. Moffatt, Canadian Coast Guard
D.S.F. Mulligan, South Africa
North Sea Directorate, Rijkswaterstaat, Netherlands
Oil Pollution Research Unit, UK
E.H. Owens, Geoscience Services, UK
D.S. Page, Bowdoin College, USA
W. Park, Mobil Oil, USA
K. Port, Department of the Environment, UK
B. Pyburn, British Petroleum, UK
A.D. Read, Exploration and Production Forum, UK
S.L. Ross, Canada
Rotortech, UK
J.K. Rudd, Amoco Europe & West Africa, UK
R. Seward, Tindall, Riley, UK
L.B. Solsberg, Canada
J. Spillman, Cranfield Institute of Technology, UK
P.S. Stamp, National Agency of Environmental Protection, Denmark
G. Teasdale, Ministry of Defence, UK
O. Terling, Sweden
N. Thelwell, Imperial Chemical Industries, UK
S. Uhler, Swedish Coast Guard
UMP Chemicals, UK
D. Voy, CB Helicopters, UK
Warren Spring Laboratory, UK
J. Wardley-Smith, UK
E.C. Wayment, Santos, Australia
J. Wonham, International Maritime Organization

Lay-out and most illustrations by Chris Goodman.